The Four Rules of Metric Measures and Time

K. A. Hesse

Longman

LONGMAN GROUP LIMITED
London
*Associated companies, branches and representatives
throughout the world*

© K. A. Hesse 1958
Metric edition © Longman Group Ltd
*All rights reserved. No part of this publication may be
reproduced, stored in a retrieval system, or transmitted
in any form or by any means, electronic, mechanical,
photocopying, recording, or otherwise, without the
prior permission of the Copyright owner.*

First published 1958
Metric edition first published 1970
Second impression 1971

Pupils' book   ISBN 0 582 18175 5
Teacher's Book ISBN 0 582 18176 3

Printed in Great Britain by
William Clowes and Sons, Limited
London, Beccles and Colchester

Complete:

Further
practice
pages

A Copy these numbers and ring the hundreds:

250	107	618	710	108	180

3 and 4

B Write in figures the value of the 4 in each of these numbers:

461	2147	4030	604

5

C Which way do we move figures to make the value less?

By how much in value does a number change when its figures are moved 2 places to the left?

6

D How many hundreds are there in each number?

562	1232	2010	8006

7 A–H

E If the figure 8 in each of these numbers is changed to figure 5, by how much is each number decreased?

182	2806	1088	8680

7 I–N

F How many tenths in 1 unit? in 4 units?

How many hundredths in 2 units? in 3 tenths?

8 and 9

G Write in words the value of the 6 in each number:

163	2·63	0·269

10

H Write how many thousandths there are in

0·012	0·009	0·03	0·106

11 and 12

I
$0·07+0·04=$ tenths and hundredths
$0·14+0·08=$ tenths and hundredths
$0·008+0·005=$ hundredths and thousandths
$0·016+0·009=$ hundredths and thousandths

13

J Add:

2·7	1·34	0·037	6·354
4·0	0·86	0·085	0·908
3·9	5·09	0·006	3·076

14

Complete:

			Further practice
			pages
A 1·3−0·8= 0·12−0·09=	2·1−0·6= 0·21−0·08=	3−0·4= 2−0·06=	*15*
B 0·6−0·08=	0·32−0·06=	1·24−0·09=	*16*

C

0·32 −0·18	0·43 −0·37	2·01 −0·29	4·03 −2·95	*17 A–M*

D

0·012 −0·004	0·034 −0·018	0·502 −0·236	5·06 −1·892	*17 N–U*

E 0·3×3=   0·5×7=   0·9×8=   *18 A–O*

F

2·4 × 6	5·7 × 8	2·16 × 7	1·92 × 12	*18 P–R*

G 0·5×0·3=   0·7×0·9=   0·8×0·5=   *19 A–D*

H

2·4 × 0·5	4·3 × 1·1	12·6 × 0·7	30·5 × 1·2	*19 E–F*

I

0·27 × 0·4	0·79 × 1·1	0·208 × 0·9	0·085 × 1·2	*19 I–K*

J 1·6÷4=   0·72÷9=   0·056÷8=   *20 A–Q*

K

4)2·08   6)8·04   9)0·774   11)3·003   *20 R–T*

L 2·4÷0·6=   3·01÷0·7=   4·4÷0·08=   *21 A–F*

Take each answer to the third decimal place:

M 0·6)2·64   0·9)3·64   0·11)0·708   0·12)1·106   *21 G–L*

Whole numbers

We use nine figures and a zero to write any number, great or small. When used alone we call them **units**.

A   Which number comes after 9?

Our numbers are made up of sets of ten, so we use the figure 1 again with a place-holder zero to show a set of ten without a unit.

B   What comes after 10?

C   What comes after 1 ten and 9 units?

Look at the grid below. Each row contains units but no sets of ten.

D   Copy the grid and put in the set of ten in the second row.

E   Put in the other tens and complete the statement after each row.

0	1	2	3	4	5	6	7	8	9	units
0	1	2	3	4	5	6	7	8	9	ten and units
0	1	2	3	4	5	6	7	8	9	tens and units
0	1	2	3	4	5	6	7	8	9	tens and units
0	1	2	3	4	5	6	7	8	9	tens and units
0	1	2	3	4	5	6	7	8	9	tens and units
0	1	2	3	4	5	6	7	8	9	tens and units
0	1	2	3	4	5	6	7	8	9	tens and units
0	1	2	3	4	5	6	7	8	9	tens and units
0	1	2	3	4	5	6	7	8	9	tens and units
0	1	2	3	4	5	6	7	8	9	tens and units
0	1	2	3	4	5	6	7	8	9	tens and units

F   State how many tens there are in

    89   109   103   114   214   126   226   478

A   You can use your grid to practise counting on from 0 to 119 or farther, from 1 to 111 in tens, from 2 to 112 in tens, etc.

B   You can use your grid to practise counting back from 119 to 0, or in tens from 110 to 0, 114 to 4, etc.

C   See what patterns you get by crossing the grid diagonally, both upwards and downwards from any set of ten.

Use your grid to answer the following:

D $12+10=$	$32+10=$	$22+20=$	$52+30=$	$22+70=$
E $33+5=$	$73+5=$	$28+60=$	$48+20=$	$66+30=$
F $72-10=$	$72-20=$	$63-20=$	$84-30=$	$95-40=$
G $89-8=$	$89-18=$	$73-8=$	$73-18=$	$85-27=$
H $100-40=$	$102-40=$	$112-40=$	$100-30=$	$106-30=$
I $110-20=$	$112-20=$	$117-30=$	$116-70=$	$123-50=$

Copy these numbers and ring the tens:

J 23	47	62	80	100	107	101	115
K 90	120	125	135	145	192	200	210
L 105	205	405	705	720	761	403	290

Copy these numbers and ring the hundreds:

M 100	200	300	110	140	150	250	104
N 126	326	506	608	120	172	370	804

Write in figures:

0   three tens and four units   five tens and seven units

P   one hundred, two tens and eight units

Q   seven hundred no tens and three units

R   twelve tens and nine units   twenty tens and no units

S   seventeen   eighty-four   one hundred and twenty-two

T   ninety   two hundred   three hundred and nine

U   twelve   a hundred and one   eight hundred and eleven

Put column headings like these: H T U

Write these numbers under them:

A   3 tens and 4 units

B   4 tens and 0 units

C   4 hundreds, 2 tens and 5 units

4 tens = 4 units × 10   4 hundreds = 4 tens × 10

D   What did you notice about the position of the figure 4 each time its value became
10 times greater in A, B and C?

It moved one place to the **left**.

Remember: each move of a figure 1 place to the left makes its value 10 times greater.

E   How much greater in value is the figure 4 when moved 2 places to the left?

We have 4 units × 10 × 10   or   4 units × 100

Remember: by moving a figure 2 places to the **left** its value is made 100 times greater.

Rewrite each of these numbers when multiplied by 10.

F	H T U	H T U	H T U	H T U	H T U	H T U
	3 4	8	4 0	6 1	6	7 0

Did you put in the place-holder 0?

G   What number equals 100 × 10?

H   What number equals 300 × 10?

When hundreds are multiplied by 10 we have a column of **thousands**.

Rewrite these numbers when multiplied by 10.

I	Th H T U	Th H T U	Th H T U	Th H T U	Th H T U
	4 3	7 4 5	2 0	3 0 1	7 8 0

Write in figures the value of the figure 7 in each of these numbers:

J	3 670	4 761	2 017	709
K	7 641	7 006	3 070	1 760

Rewrite these numbers when multiplied by 100:

A	Th H T U	Th H T U	Th H T U	Th H T U	Th H T U
	3	3 1	5 0	8	7 9

Write these numbers in words:

B   273   406

C   2 730   460

D   2 703   416

E   2 073   604

F   What does 3 become when multiplied by 10?

G   What does 3 become when multiplied by 100?

H   What does 300 become when divided by 100?

I   What does 30 become when divided by 10?

J   What does 300 become when divided by 10?

K   What does 340 become when divided by 10?

L   Which way do we move figures to make their value greater?

M   Which way do we move figures to make their value less?

N   How much greater does the value of a figure become when it is moved one place to the left?

O   How much less does the value of a figure become when it is moved two places to the right?

P   How much less does the value of a figure become when it is moved one place to the right?

Q   How much greater does the value of a figure become when it is moved two places to the left?

R   When multiplying a number by 100 how many place-holder zeros do you need?

S   When multiplying a number by 1 000 how many place-holder zeros do you need?

Large numbers

How many tens are there in each of these numbers?

A   100   200   110   114   314   406   730

B   206   678   920   1 610   1 000   2 470   5 060

Add 20 to each of these numbers (note that 20=2 tens)

C   236   470   380   786   493   6 061   5 291

Add 40 to each of these numbers:

D   59   61   87   276   589   4 793   7 080

Take 10 from each of these numbers:

E   43   79   130   100   605   1 000   2 100

Take 30 from each of these numbers:

F   74   152   631   214   5 629   1 010   3 037

How many hundreds are there in each of these numbers?

G   571   1 260   2 050   807   7 060   8 900

H   95   2 090   17 600   21 209   8 006   10 095

Add 300 to each of these numbers:

I   507   751   907   1 293   2 800   4 976

Take 400 from each of these numbers:

J   781   906   1 081   1 481   3 060   2 281

K   1 169   3 073   13 067   22 901   14 170   40 000

How many thousands are there in each of these numbers?

L   3 705   13 705   30 800   10 701   15 600   9 085

If the figure 3 in each of these numbers is changed to figure 5 by how much is each number increased?

M   234   1 306   713   2 030   13 906   25 390

If the figure 7 in each of these numbers is changed to figure 4 by how much is each number decreased?

N   307   1 760   17 800   2 770   877   7 870

1*

A   What happens when we move this   H T U
number one place to the right?   2 4 6

B   If 2 people share 3 apples, how much does each get?

C   Write the fraction represented by the shaded part in each shape:

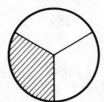 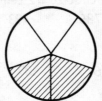 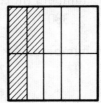 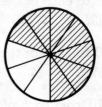

On your grid you saw that our numbers proceed in groups of ten.
A figure moved one place to the left is ten times greater.
A figure moved one place to the right is ten times less.
In the shapes above the first circle is divided into 3 equal parts.

D   What do we call each part?

E   What do we call each part of the second circle?

F   What part of the second circle is shaded?

G   What do we call each part of the third circle?

H   What part of the third circle is shaded?

I   What part of the third circle is not shaded?

J   What part of the second square is not shaded?

Here are three apples.

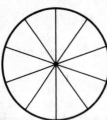 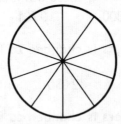 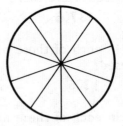

K   Into how many parts is each apple divided?

L   If ten boys share the apples equally how much will each get?

M   If ten boys shared 7 apples equally how much would each get?

When we divide by 10 we have tenths.

1 unit divided by $10 = \frac{1}{10}$ 6 units divided by $10 = \frac{6}{10}$

A What happens to any number when moved one place to the right?

B What does 246 become when moved one place to the right?

$$
\begin{array}{c}
\text{H T U} \\
2\ 4\ 6 \div 10
\end{array}
=
\begin{array}{c}
\text{H T U} \\
2\ 4
\end{array}
$$

We can put remainders without the fraction.

Our numbers go up in groups of ten and down in groups of ten.

$$
\begin{array}{c}
\text{H T U} \\
2\ 4\ 6 \div 10
\end{array}
=
\begin{array}{c}
\text{H T U t} \\
2\ 4\ 6
\end{array}
\quad \text{(t stands for tenths)}
$$

To show that a figure has a value less than a unit we put in a dot after the unit. We call the dot a **decimal point**.

$$
246 \div 10 =
\begin{array}{c}
\text{T U} \cdot \text{t} \\
2\ 4 \cdot 6
\end{array}
\quad \text{(say, 'Twenty-four point six.')}
$$

What is the value of the figure 7 in each of these numbers?

C   1·7   30·7   17·5   72·6

Note that if there are no units we write 0·5 and 0·6.

Here is a unit. Next to it is the same unit divided into tenths and next to that is the same unit divided yet again ten times.

 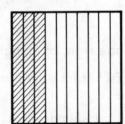 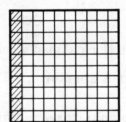

D What do we call the shaded parts in the second shape?

E How many parts are there in the third shape?

F What do we call the parts in the third shape?

G How many tenths in 1 unit?   2 units?   5 units?

H How many hundredths in a unit?   2 units?   4 units?

I How many hundredths in one tenth?   3 tenths?   7 tenths?

We can write another column:

Th H T U · t h   (h for hundredths)
1 3 7 6 · 2 4

A Write how many tenths there are altogether in

0·3   1·3   1·7   2·4   0·8   0·6   4·5

B Write how many hundredths there are altogether in

0·07   0·17   1·17   1·06   0·58   3·09   2·14

C Write as decimals. Remember the place-holder zeros.

$\frac{1}{10}$ $\frac{1}{100}$ $\frac{3}{100}$ $\frac{7}{10}$ $\frac{13}{100}$ $\frac{87}{100}$ $1\frac{9}{10}$

Complete:

D Moving figures one place to the right makes them   times less.

E Moving tens one place to the right makes them

F Moving units one place to the right makes them

G Moving tenths one place to the right makes them

H Moving units two places to the right makes them

Write in words the value of the figure 9 in each of these numbers:

I   92   0·9   9·45

J   7·98   3·19   10·49

K   What do you think should come to the right of hundredths?

Our number columns are now: Th H T U · t h th   (th for thousandths).

Write in figures:

L   four tens, no units, three tenths and two hundredths.

M   one unit, four tenths, six hundredths and two thousandths.

N   no units, three tenths, no hundredths and nine thousandths.

O   four hundredths and seven thousandths.

P   eight units, three hundredths and one thousandth.

Q   five tenths and eight thousandths.

Divide each of these numbers by 10:

R   540   16·7   40·5   7   19   80

S   0·6   0·62   3·71   0·24   1·02   0·09

Complete:

A   23 =   tens and   units   or   units

B   40 =   tens and   units   or   units.

C   18 =   ten and   units   or   units.

D   90 =   tens and   units   or   units.

E   134 =   hundred and   tens and   units
               or   tens and   units   or   units.

F   116 =   hundred and   ten and   units
               or   tens and   units   or   units.

G   212 =   hundreds and   ten and   units   or   tens and   units.

H   436 =   tens and   units   or   units.

I   358 =   units   or   tens and   units.

J   529 =   units   or   tens and   units   or   hundreds and   units.

K   706 =   hundreds and   units   or   tens and   units   or   units.

L   410 =   hundreds and   units   or   tens and   units   or   units.

M 1112 =   thousand and   hundred and   ten and   units
               or   hundreds and   ten and   units   or   tens and   units.

N 2134 =   thousands and   hundred and   tens and   units
               or   hundreds and   tens and   units   or   tens and   units.

O 4728 =   hundreds and   tens and   units   or   tens and   units.

P 6132 =   hundreds and   tens and   units   or   units.

Q 5906 =   hundreds and   tens and   units   or   tens and   units.

R 7083 =   tens and   units   or   hundreds and   units.

S 6400 =   hundreds and   tens and   units   or   tens and   units.

T 3089 =   hundreds and   units   or   tens and   units   or   units.

U   760 =   tens and   units   or   hundreds and   units.

V 2090 =   hundreds and   units   or   tens and   units.

Complete:

A   Moving figures one place to the right makes them   times less.

B   Moving figures two places to the right makes them   times less.

C   Moving figures three places to the right makes them   times less.

D   Moving figures one place to the left makes them 10 times

E   Moving figures two places to the left makes them   times greater.

F   Moving figures three places to the left makes them   times greater.

Multiply each of these numbers by 10:

G   14   230   2·3   7·8   0·4   0·42   0·731

Multiply each of these numbers by 100:

H   7   1·2   17·6   0·8   0·08   1·376   0·084

Divide each of these numbers by 10:

I   60   3   0·7   0·76   5·01   30·9   0·08

Divide each of these numbers by 100:

J   3 700   408   90   6   2·4   10·5   0·4

Multiply each of these numbers by 1 000:

K   72   4·7   40·8   2·67   0·58

L   1·302   0·64   1·072   0·034   0·009

Divide each of these numbers by 1 000:

M   8 000   1 706   451   301   76

N   600   9   30   4   18

O   7 439   186   5   76   9

Write how many thousandths there are in

P   0·007   0·016   0·106   0·217   0·408

Re-arrange in order of size, smallest first:

Q   1·078, 0·746, 1·7, 0·197, 1·079

R   1·049, 8·2, 5, 1·068, 0·987

Complete:

A   1·1=   unit and   tenth or   tenths   1·2=   tenths

B   2·1=   units and   tenth or   tenths   2·4=   tenths

C   1·6=   unit and   tenths or   tenths   1·3=   tenths

D   3·8=   units and   tenths or   tenths   3·5=   tenths

E   2·9=   tenths   8·2=   tenths   7·9=   tenths

F   0·7+0·8=   tenths=   unit and   tenths=

G   0·9+0·8=   tenths=   unit and   tenths=

H   0·11=   tenths and   hundredths or   hundredths.

I   0·23=   tenths and   hundredths or   hundredths.

J   0·71=   tenths and   hundredths or   hundredths.

K   0·69=   tenths and   hundredths or   hundredths.

L   0·08=   tenths and   hundredths or   hundredths.

M   0·06+0·05=   hundredths or   tenths and   hundredths=

N   0·05+0·08=   hundredths or   tenths and   hundredths=

O   0·09+0·04=   hundredths or   tenths and   hundredths=

P   0·17+0·05=   hundredths or   tenths and   hundredths=

Q   0·216=   tenths and   hundredths and   thousandths
        or   hundredths and   thousandths or   thousandths.

R   0·347=   hundredths and   thousandths or   thousandths.

S   0·064=   tenths and   hundredths and   thousandths
        or   hundredths and   thousandths or   thousandths.

T   0·078=   tenths and   hundredths and   thousandths
        or   hundredths and   thousandths or   thousandths.

U   0·039=   hundredths and   thousandths or   thousandths.

V   0·006=   hundredths and   thousandths or   thousandths.

W   0·007+0·004=   thousandths or   hundredths and   thousandths
        =

X   0·008+0·006=   thousandths or   hundredths and   thousandths
        =

A   How many units equal one ten?

B   How many hundreds equal one thousand?

C   How many tenths equal one unit?

D   How many hundredths equal one tenth?

E   How many thousandths equal one hundredth?

Complete:

F   8 units+3 units=   units=1   and 1

G   8 units+5 units=   units=1   +3

H   8 tenths+5 tenths=   tenths=1   +3

I   0·8+0·5=   0·8+0·7=   0·6+0·7=   0·6+0·9=

J   0·7+0·8=   0·8+0·6=   0·7+0·7=   0·8+0·8=

K

1·2	1·3	2·2	1·3	3·4	2·5	1·7	3·6
3·5	2·4	1·6	0·5	0·3	1·4	2·3	2·0
2·1	1·5	2·3	2·5	1·6	0·6	0·5	1·9

L   7 hundredths+5 hundredths=   hundredths=1   +2

M   0·07+0·05=   0·08+0·03=   0·07+0·06=

N   0·06+0·06=   0·09+0·04=   0·06+0·08=

O   0·05+0·08=   0·06+0·07=   0·09+0·08=

P

0·03	0·04	0·13	0·24	0·16	0·34	1·25
0·04	0·05	0·23	0·05	0·02	0·26	2·43
0·04	0·04	0·16	0·36	0·53	0·55	1·66

Q   6 thousandths+6 thousandths=1   +2

R   0·008+0·003=   0·009+0·004=   0·008+0·007=

S   0·006+0·007=   0·006+0·009=   0·007+0·009=

T

0·005	0·004	0·007	0·014	0·125	1·241
0·001	0·003	0·008	0·032	0·234	2·608
0·007	0·008	0·006	0·018	0·156	2·372

Complete:

A   15−9=   ten and   units−   units=   units−   units=

B   23−9=   tens and   units−   units
         = 1 ten and   units−   units=1 ten and   units=

C   45−8=   tens and   units−   units
         =   tens and   units−   units=   tens and   units=

D   1·2=   unit and   tenths=   tenths   3·1=   tenths

E   2·3=1 unit and   tenths   5·7=4   and   tenths

F   1·3−0·8=   unit and   tenths−8   =   tenths−   tenths=

G   1·5−0·6= 1   and 5   −6   =   tenths−   tenths=

H   1·4−0·7=   1·2−0·9=   1·3−0·6=   1·6−0·8=

I   3·2−0·9= 3   and 2   −9
         = 2   and 1   and 2   −9
         = 2   and   tenths−   tenths
         = 2   and   tenths=

J   2·4−0·8=   units and   unit and   tenths−   tenths
         =   units and   tenths−   tenths=

K   5·3−0·6=   units and   unit and   tenths−   tenths
         =   units and   tenths−   tenths=

L   4·5−0·7= 3 units and   tenths−7 tenths=

M   3·4−0·9=   units and   tenths−   tenths=

N   6·1−0·8=   units and   −   =

O   2·3−0·9=   4·1−0·8=   7·4−0·7=   5·6−0·9=

P   0·14−0·06= 1   and 4   −6
         =   hundredths−   hundredths=

Q   0·15−0·08=   hundredths−   hundredths=

R   0·13−0·06=   hundredths−   hundredths=

S   0·14−0·09=   hundredths−   hundredths=

T   0·13−0·08=   0·26−0·09=   0·35−0·06=

2*

Complete:

A   0·15—0·07=   0·14—0·08=   0·17—0·09=

B   0·32—0·06=   tenths and   hundredths—   hundredths
         = 2 tenths and 1   and 2   —6
         = 2   and   hundredths—6
         =   tenths and   hundredths=

C   0·32—0·07= 3   and 2   —7
         = 2 tenths and   hundredths—7
         =   tenths and   hundredths=

D   0·53—0·09= 5   and   —9
         =   tenths and   hundredths—   hundredths
         =   tenths and   hundredths=

E   0·31—0·06=   0·52—0·08=   0·75—0·07=

F   0·4—0·07= 4   —7
         = 3 tenths and   hundredths—   hundredths
         =   tenths and   hundredths=

G   0·7—0·05= 7   —5
         =   tenths and   hundredths—   hundredths
         =   tenths and   hundredths=

H   0·6—0·08=   tenths and   hundredths—   hundredths
         =   tenths and   hundredths=

I   0·8—0·03=   tenths and   hundredths—   hundredths
         =   tenths and   hundredths=

J   0·7—0·06=   0·8—0·08=   0·3—0·07=

K   0·51—0·07=   0·36—0·09=   0·6—0·04=

L   1·56—0·04=   1·38—0·06=   1·42—0·08=

M   1·34—0·07=   1·2—0·05=   1·3—0·07=

N   1·5—0·06=   1·7—0·09=   1·5—0·08=

Subtraction

A 0·8 − 0·6 =   0·09 − 0·05 =   0·007 − 0·004 =

B 1·3 =   tenths   1·3 − 0·6 =

C 1·2 − 0·6 =   1·4 − 0·8 =   1·6 − 0·9 =   2·1 − 0·8 =

D 1·5 − 0·8 =   2·5 − 0·8 =   3·2 − 0·7 =   4·3 − 0·7 =

E

1·4	1·3	1·6	1·5	2·2	2·3	3·1	4·3
−0·6	−0·8	−0·9	−0·7	−0·8	−0·9	−1·8	−1·6

F 0·14 =   hundredths   0·14 − 0·08 =

G 0·12 − 0·09 =   0·15 − 0·08 =   0·17 − 0·09 =

H 0·21 − 0·06 =   0·24 − 0·07 =   0·35 − 0·08 =

I

0·13	0·17	0·16	0·23	0·25	0·34	0·55
−0·06	−0·09	−0·08	−0·07	−0·08	−0·07	−0·09

J

0·34	0·56	0·22	0·24	2·67	2·43	3·84
−0·16	−0·18	−0·18	−0·18	−1·18	−1·27	−0·46

K 0·012 =   thousandths   0·018 =   thousandths

L 0·012 − 0·008 =   0·015 − 0·007 =   0·014 − 0·008 =

M 0·015 − 0·009 =   0·017 − 0·009 =   0·015 − 0·009 =

N

0·014	0·016	0·013	0·035	0·027	0·042
−0·006	−0·008	−0·006	−0·016	−0·008	−0·015

O

0·034	0·051	0·073	0·131	0·523	0·743
−0·016	−0·028	−0·065	−0·016	−0·219	−0·218

P

0·103	1·025	1·002	2·08	1·013	3·05
−0·046	−0·407	−0·078	−0·396	−0·996	−1·909

Write in figures. Work across the page:

A	fifteen units	fifteen tenths	fifteen hundredths
B	12 tenths	27 tenths 20 tenths	50 tenths
C	12 hundredths	17 hundredths	27 hundredths
D	60 hundredths	31 hundredths	80 hundredths
E	132 hundredths	207 hundredths	430 hundredths
F	8 hundredths	10 hundredths	5 hundredths
G	132 thousandths	503 thousandths	67 thousandths
H	91 thousandths	90 thousandths	8 thousandths
I	60 thousandths	4 thousandths	9 thousandths

Multiply each of these numbers by 10:

J	2·5	0·72	0·8	1·73	1·04	0·038

Multiply each of these numbers by 100:

K	1·76	0·35	4·07	0·06	0·042	0·009
L	0·03	0·6	0·8	1·07	0·206	0·061

Complete. Work across the page:

M	$2\times2=$	$0·2\times2=$	$0·2\times6=$	$0·7\times4=$
N	$3\times4=$	$0·3\times4=$	$0·5\times7=$	$0·6\times5=$
O	$0·5\times8=$	$0·6\times7=$	$0·8\times5=$	$0·9\times7=$

P

$$
\begin{array}{cccccc}
3·2 & 3·4 & 2·8 & 4·6 & 3·7 & 4·6 \\
\times\ 4 & \times\ 3 & \times\ 7 & \times\ 6 & \times\ 9 & \times\ 9 \\
\hline
\end{array}
$$

Q

$$
\begin{array}{cccccc}
2·13 & 1·68 & 3·07 & 2·08 & 1·49 & 3·09 \\
\times\ 7 & \times\ 6 & \times\ 5 & \times\ 9 & \times\ 12 & \times\ 12 \\
\hline
\end{array}
$$

R

$$
\begin{array}{cccccc}
1·5 & 3·5 & 4·8 & 1·75 & 4·05 & 1·95 \\
\times\ 4 & \times\ 6 & \times\ 5 & \times\ 12 & \times\ 8 & \times\ 12 \\
\hline
\end{array}
$$

Study:

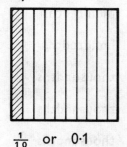 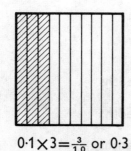 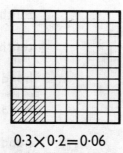 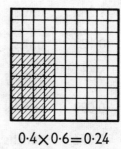

$\frac{1}{10}$ or 0·1   0·1×3=$\frac{3}{10}$ or 0·3   0·3×0·2=0·06   0·4×0·6=0·24

Write answers only. Work across the page:

A	0·2×3 =	0·2×0·3=	0·1×0·7=	0·2×0·4=
B	0·3×0·3=	0·3×0·4=	0·5×0·5=	0·4×0·7=
C	0·6×0·7=	0·6×0·5=	0·5×0·8=	0·8×0·9=
D	0·7×1·1=	0·8×1·2=	0·5×1·2=	0·9×1·2=

Multiply:

E	1·3 ×0·4	1·5 ×0·3	2·7 ×0·06	3·5 ×0·7	6·4 ×0·9	5·8 ×1·1

F	13·2 ×0·6	14·6 × 1·1	12·5 ×0·6	20·8 ×0·5	30·5 ×0·8	40·5 ×1·2

Write answers only. Work across the page:

G	0·02×4=	0·02×0·4=	0·03×0·2=	0·11×0·3=
H	0·12×0·3=	0·11×0·6=	0·12×0·5=	0·12×0·9=

Multiply:

I	0·09 × 0·3	0·18 × 0·4	0·36 × 0·7	0·56 × 0·5	0·75 × 0·8	0·89 × 1·1

J	0·36 × 0·5	1·28 × 0·7	3·17 × 0·6	1·08 × 0·9	2·05 ×0·08	3·05 × 1·2

K	0·312 × 0·4	0·206 × 0·7	0·604 × 0·9	2·086 × 1·2	0·079 × 1·1	0·095 × 1·2

State how many:

A	1·2	units		tenths	hundredths
B	0·12	units		tenths	hundredths
C	0·37	tenths		hundredths	thousandths
D	0·083	tenths		hundredths	thousandths
E	0·076	tenths		hundredths	thousandths
F	0·305	tenths		hundredths	thousandths
G	0·42	tenths		hundredths	thousandths
H	2·006	tenths		hundredths	thousandths
I	0·008	tenths		hundredths	thousandths

Study this shape.

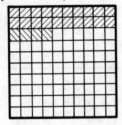

J   How many rows of small squares are there?

K   How many small squares in the shape?

L   What part of the shape is shaded?

M   Divide that number by 4.

N   Write what you did in **M** as a number sentence or equation in fractions and in decimals.

Write answers only:

O   12 ÷4=   1·2÷4 =   1·5÷3 =   2·8÷4 =

P   0·32÷4=   0·44÷11=   0·36÷9 =   0·72÷8 =

Q   0·42÷7=   0·08÷4 =   0·024÷6=   0·048÷6=

Divide:

R

2)4·6   2)0·46   3)0·639   4)1·68   5)2·35

S   6)7·02   8)4·024   7)3·563   9)9·72   8)0·464

T   5)2·005   7)0·469   9)1·008   11)1·001   12)1·104

Take answers to the 3rd decimal place:

A $5\overline{)0{\cdot}064}$ $7\overline{)1{\cdot}025}$ $6\overline{)0{\cdot}52}$ $8\overline{)1{\cdot}7}$ $11\overline{)3{\cdot}03}$

B $7\overline{)0{\cdot}8}$ $8\overline{)3{\cdot}7}$ $11\overline{)50}$ $9\overline{)9{\cdot}65}$ $12\overline{)10{\cdot}9}$

$$0{\cdot}3 \div 3 = \tfrac{3}{10} \div 3 = \tfrac{1}{10} = 0{\cdot}1 \qquad\qquad 0{\cdot}3 \div 0{\cdot}3 = \tfrac{3}{10} \div \tfrac{3}{10} = 1$$

C $0{\cdot}8 \div 0{\cdot}2 =$ $0{\cdot}8 \div 0{\cdot}4 =$ $0{\cdot}9 \div 0{\cdot}3 =$ $0{\cdot}6 \div 0{\cdot}3 =$

D $1{\cdot}2 \div 0{\cdot}3 =$ $1{\cdot}2 \div 0{\cdot}4 =$ $1{\cdot}5 \div 0{\cdot}5 =$ $1{\cdot}8 \div 0{\cdot}6 =$

$$0{\cdot}3 \div 0{\cdot}03 = \tfrac{3}{10} \div \tfrac{3}{100} = \tfrac{30}{100} \div \tfrac{3}{100} = 10$$

E $0{\cdot}8 \div 0{\cdot}02 =$ $0{\cdot}9 \div 0{\cdot}03 =$ $1{\cdot}8 \div 0{\cdot}6 =$ $2{\cdot}4 \div 0{\cdot}08 =$

F $2{\cdot}1 \div 0{\cdot}07 =$ $0{\cdot}3 \div 0{\cdot}06 =$ $0{\cdot}4 \div 0{\cdot}08 =$ $3{\cdot}6 \div 0{\cdot}09 =$

Multiply each number in these pairs by 10:
G 2·1 and 0·6   2·73 and 3·4

Multiply each number in these pairs by 100:
H 3·6 and 0·04   7·012 and 0·11

Take each answer to the fourth decimal place:

I $0{\cdot}2\overline{)0{\cdot}46}$ $0{\cdot}3\overline{)0{\cdot}97}$ $0{\cdot}6\overline{)2{\cdot}15}$ $0{\cdot}5\overline{)3{\cdot}108}$ $0{\cdot}4\overline{)9{\cdot}03}$

J $0{\cdot}8\overline{)0{\cdot}103}$ $0{\cdot}7\overline{)2{\cdot}06}$ $0{\cdot}8\overline{)0{\cdot}304}$ $0{\cdot}9\overline{)5{\cdot}46}$ $1{\cdot}1\overline{)6{\cdot}03}$

K $0{\cdot}06\overline{)4{\cdot}302}$ $0{\cdot}07\overline{)4{\cdot}903}$ $0{\cdot}09\overline{)0{\cdot}13}$ $0{\cdot}11\overline{)2{\cdot}607}$ $0{\cdot}08\overline{)6{\cdot}10}$

L $0{\cdot}07\overline{)0{\cdot}7201}$ $0{\cdot}12\overline{)0{\cdot}3722}$ $0{\cdot}11\overline{)2{\cdot}005}$ $0{\cdot}09\overline{)1{\cdot}06}$ $0{\cdot}12\overline{)6{\cdot}10}$

A   Look at a metrestick or tape and at a ruler.

Write m for metres or cm for centimetres to show which measure you would use (to the nearest metre or centimetre) to find the following measurements:

B   height of a bottle

circumference of a bottle

Mark's height

Mark's chest

C   Angela's height   Angela's waist

D   Paul's span   Paul's pace

E   length of a room

width of a room

length of a girder

width of a girder

F   Length of a corridor   length of a hall

G   length of a bridge

length of a street

height of a lamp post

distance between lamp posts

H   Jennifer's full stretch finger tip to finger tip.

I   Jennifer's full reach, tip toe to finger tip.

J   Make a list of three measurements in or near your school which you would make in metres and another three items you would make in centimetres.

Complete:

								Further practice pages
A	12 mm= cm		25 mm= cm		47 mm= cm			24 A–N
B	114 mm= cm		106 mm= cm		290 mm= cm			24 O–U
C	1·4 cm= mm		10 cm= mm		6·7 cm= mm			25 A–O
D	114 cm= m		206 cm= m		600 cm= m			26 A–Q
E	2 m 16 cm= m		7 m 30 cm= m		20 m 85 cm= m			26 R–V
F	State these lengths to the nearest centimetre: 3 cm 6 mm 2 cm 5 mm 73 mm							27 A–L
G	State these lengths to the nearest 10 centimetres: 2 m 25 cm 4 m 62 cm 6 m 5 cm							27 M–S
H	1 m 12 cm 5 mm= m 2 m 47 cm 6 mm= m 2 m 30 cm 8 mm= m 1 m 6 cm 2 mm= m							28 A–U
I	1·18 m= m cm 1·213 m= m cm mm 0·306 m= m cm mm 0·076 m= m cm mm							29 A–L
J	State these measurements in metres to the nearest centimetre: 1·478 m 0·304 m 1·097 m							29 M–U
K	276 m= km 1 km 270 m= km 2 km 90 m= km							30 A–U
L	2·08 km= m 0·067 km= m 1·3 km= m							31 A–W
M	State these measurements to the nearest kilometre: 100·64 km 67·95 km 209·712 km							32 A–M
N	State these measurements in kilometres to the nearest 10 metres: 1·645 km 0·995 km 10·096 km							32 N–R

3*

cent as the first part of any word usually means **one hundred**.

A   Look in your dictionary for these words and write their meanings:

century   cent   centenary   centurion   centigrade

milli as the first part of a word usually means **one thousand**.

B   Look in your dictionary for these words: millipede   millennium

C   How many centimetres equal one metre?

D   How many millimetres equal one metre?

E   How many millimetres equal one centimetre?

Complete:

F   10 mm=   cm   1 mm= $\frac{1}{}$ cm $\frac{1}{10}$ cm=0· cm   1 mm= 0· cm

G   1 cm 1 mm=1 cm+$\frac{1}{}$cm=1· cm   3 mm=   cm   7 mm=   cm

H   1 cm 2 mm= 1· cm   1 cm 4 mm=   cm   2 cm 4 mm=   cm

I   1 cm 7 mm=   cm   2 cm 3 mm=   cm   2 cm 5 mm=   cm

J   3 cm 6 mm=   cm   5 cm 1 mm=   cm   8 cm 7 mm=   cm

K   4 cm 3 mm=   cm   6 cm 9 mm=   cm   9 cm 8 mm=   cm

L   15 mm=   cm   mm=   cm   17 mm=   cm   mm=   cm

M   23 mm=   cm   mm=   cm   26 mm=   cm   mm=   cm

N   37 mm=   cm   mm=   cm   52 mm=   cm   mm=   cm

O   To change millimetres to centimetres we   by ten.

P   To divide numbers by 10 we move figures   place to the   .

Q   25 mm=   cm   36 mm=   cm   49 mm=   cm   70 mm=   cm

R   39 mm=   cm   50 mm=   cm   87 mm=   cm   69 mm=   cm

S   114 mm=   cm   106 mm=   cm   93 mm=   cm   149 mm=   cm

T   205 mm=   cm   80 mm=   cm   109 mm=   cm   300 mm=   cm

U   450 mm=   cm   9 mm=   cm   500 mm=   cm   70 mm=   cm

Here are parts of some rulers. Write the length of each part, first in millimetres and then in centimetres:

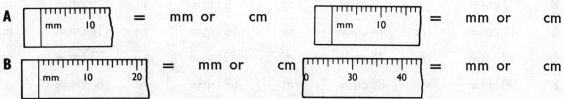

A = mm or cm   = mm or cm

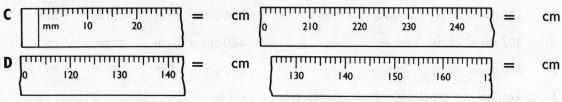

B = mm or cm   = mm or cm

Write these lengths in centimetres only:

C = cm   = cm

D = cm   = cm

E   To change centimetres to millimetres we   by ten.

F   To multiply numbers by 10 we move the figures   place to the   .

Complete:

G   1·3 cm =   mm   or   cm   mm   2·1 cm =   mm   or   cm   mm

H   2·7 cm =   mm   or   cm   mm   3·5 cm =   mm   or   cm   mm

I   5·6 cm =   mm   or   cm   mm   7·8 cm =   mm   or   cm   mm

J   4 cm =   mm   or   cm   mm   8 cm =   mm   or   cm   mm

K   9·4 cm =   mm   or   cm   mm   10·3 cm =   mm   or   cm   mm

Find the following measurements, to the nearest millimetre, stating them first in millimetres or in centimetres and millimetres and then in centimetres only:

L   The length of your pencil.

M   The length of a milk straw.

N   The thickness of a milk straw.

O   The measurements of a matchbox:

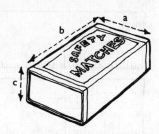

     a. the width of the box

     b. the length of the box

     c. the depth of the box.

A   100 cm =   m   1 cm = $\frac{1}{\quad}$ m $\frac{1}{100}$ = 0·   1 cm =   m

B   2 cm =   m   3 cm =   m   5 cm =   m   9 cm =   m

C   11 cm =   m   14 cm =   m   16 cm =   m   18 cm =   m

D   22 cm =   m   35 cm =   m   40 cm =   m   73 cm =   m

E   90 cm =   m   88 cm =   m   67 cm =   m   80 cm =   m

F   101 cm =   m   cm =   m   108 cm =   m   cm =   m

G   125 cm =   m   cm =   m   234 cm =   m   cm =   m

H   307 cm =   m   cm =   m   460 cm =   m   cm =   m

I   680 cm =   m   cm =   m   810 cm =   m   cm =   m

J   To change centimetres to metres we   by   .

K   To divide by 100 we move the figures   places to the   .

L   126 cm =   m   180 cm =   m   207 cm =   m   803 cm =   m

M   210 cm =   m   506 cm =   m   300 cm =   m   700 cm =   m

N   319 cm =   m   590 cm =   m   80 cm =   m   106 cm =   m

O   7 cm =   m   406 cm =   m   28 cm =   m   90 cm =   m

P   600 cm =   m   60 cm =   m   915 cm =   m   708 cm =   m

Q   19 cm =   m   230 cm =   m   307 cm =   m   800 cm =   m

Look carefully at the measurement shown on this portion of tape:

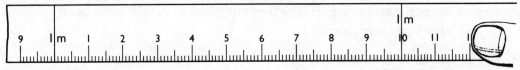

R   State the full measurement up to where the thumb is held, both in metres and centimetres and then in metres only.

Do the same for each of these:

S

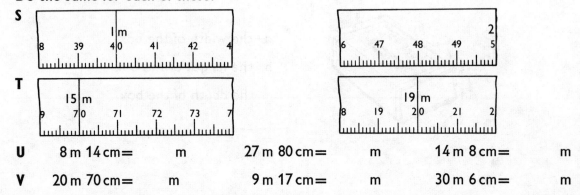

T

U   8 m 14 cm =   m   27 m 80 cm =   m   14 m 8 cm =   m

V   20 m 70 cm =   m   9 m 17 cm =   m   30 m 6 cm =   m

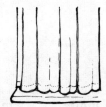

If curtains are to hang above a window sill they must be cut and sewn exactly to the right length. To hang below the sill the length need not be so exact.

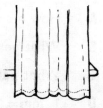

State the lengths of these lines to the nearest millimetre:

A _____ _____

B _____ _____

C _____

D _____

State the lengths of these lines to the nearest centimetre:

E _____ _____ _____

F _____ _____

G _____

H _____

State these lengths complete to the nearest centimetre:

I	2 cm 5 mm	1 cm 7 mm	3 cm 9 mm	5 cm 3 mm
	2 cm 8 mm			
J	4 cm 6 mm	6 cm 1 mm	8 cm 5 mm	7 cm 4 mm
	9 cm 7 mm			
K	6 cm 3 mm	5 cm 4 mm	11 cm 3 mm	4 cm 6 mm
	3 cm 5 mm			
L	15 mm	27 mm	40 mm	56 mm
	73 mm			

State these lengths complete to the nearest metre:

| **M** | 5 m 50 cm | 10 m 62 cm | 8 m 37 cm | 11 m 48 cm |
| **N** | 28 m 17 cm | 30 m 48 cm | 51 m 80 cm | 45 m 8 cm |

State these lengths in metres, but complete to the nearest 10 centimetres:

O	3 m 15 cm	5 62 cm	4 m 78 cm	6 m 49 cm
P	8 m 30 cm	10 m 83 cm	15 m 39 cm	20 m 28 cm
Q	21 m 7 cm	40 m 56 cm	50 m 4 cm	37 m 44 cm
R	60 m 85 cm	29 m 95 cm	30 m 97 cm	19 m 93 cm

A   How many millimetres equal one metre?

B   What part is one millimetre of one metre as a fraction?

C   What part is one millimetre of one metre as a decimal?

D   1 mm= 0·   m   2 mm= 0·   m   3 mm= 0·   m   7 mm=   m

E   10 mm=   m   12 mm=   m   14 mm=   m   18 mm=   m

F   20 mm=   m   28 mm=   m   37 mm=   m   43 mm=   m

G   83 mm=   m   60 mm=   m   79 mm=   m   91 mm=   m

H   42 mm=   cm   mm=   m   83 mm=   cm   mm=   m

I   73 mm=   cm   mm=   m   90 mm=   cm   mm=   m

J   68 mm=   cm   mm=   m   97 mm=   cm   mm=   m

K   1 m 13 cm 5 mm=   m   1 m 26 cm 7 mm=   m

L   1 m 70 cm 6 mm=   m   2 m 55 cm 0 mm=   m

State to the nearest millimetre the measurement in metres shown on each piece of tape:

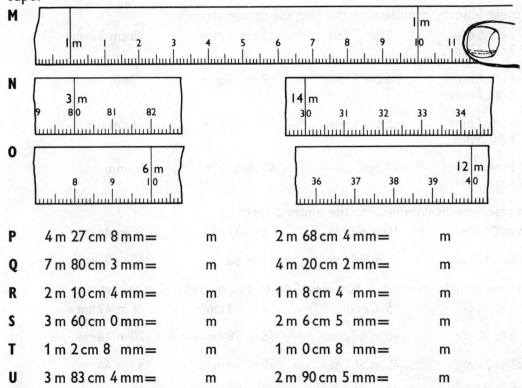

P   4 m 27 cm 8 mm=   m   2 m 68 cm 4 mm=   m

Q   7 m 80 cm 3 mm=   m   4 m 20 cm 2 mm=   m

R   2 m 10 cm 4 mm=   m   1 m 8 cm 4 mm=   m

S   3 m 60 cm 0 mm=   m   2 m 6 cm 5 mm=   m

T   1 m 2 cm 8 mm=   m   1 m 0 cm 8 mm=   m

U   3 m 83 cm 4 mm=   m   2 m 90 cm 5 mm=   m

State how you would read these amounts on a measuring tape:

A　2·7 cm=　cm　mm　　3·2 cm=　cm　mm　　4·7 cm=　cm　mm

B　1·9 cm=　cm　mm　　6·0 cm=　cm　mm　　0·8 cm=　cm　mm

C　0·9 cm=　cm　mm　　2·65 cm=　cm　mm　　1·75 cm=　cm　mm

D　1·375 m=　m　cm　mm　　　2·856 m=　m　cm　mm

E　1·806 m=　m　cm　mm　　　1·603 m=　m　cm　mm

F　2·86 m =　m　cm　mm　　　0·892 m=　m　cm　mm

G　1·076 m=　m　cm　mm　　　1·086 m=　m　cm　mm

H　2·064 m=　m　cm　mm　　　0·807 m=　m　cm　mm

I　0·908 m=　m　cm　mm　　　2·090 m=　m　cm　mm

J　3·086 m=　m　cm　mm　　　0·087 m=　m　cm　mm

K　0·605 m=　m　cm　mm　　　0·76 m =　m　cm　mm

L　1·5 m =　m　cm　mm　　　0·25 m =　m　cm　mm

How many whole metres are there in

M　3·786 m　　10·6 m　　1·083 m　　20·96 m　　0·801 m　　1·07 m

State these measurements to the nearest metre:

N　2·5 m　　2·506 m　　7·098 m　　3·496　　5·15 m　　0·806 m

O　3·05 m　　19·83 m　　21·05 m　　8·501 m　　0·651 m　　9·397 m

How many whole centimetres are there in

P　0·086 m　　0·186 m　　1·186 m　　2·158 m　　1·062 m　　3·254 m

Q　0·283 m　　0·15 m　　0·37 m　　1·37 m　　0·217 m　　0·602 m

R　1·307 m　　1·12 m　　2·1 m　　0·106 m　　0·3 m　　0·12 m

State these measurements in metres to the nearest centimetre:

S　0·125 m　　0·376 m　　0·372 m　　1·467 m　　1·464 m

T　2·349 m　　1·068 m　　1·095 m　　1·599 m　　2·098 m

U　1·903 m　　0·907 m　　0·992 m　　0·996 m　　1·998 m

Against these distances or lengths put km or m to show which part of the table of measurement you would use to measure them.

A   The length of the school playground

B   The length of the school corridor

C   The length of a river

D   The width of a river

E   The distance between London and Dover

F   The distance you could throw a cricket ball

1 000 metres=1 kilometre

G   1 metre=   km   3 metres=   km   12 m=   km

H   To divide by one thousand we move figures   places to the   .

Complete:

I   1 m= 0·   km   5 m= 0·   km   7 m= 0·   km

J   10 m= 0·   km   12 m= 0·   km   24 m= 0·   km

K   30 m= 0·   km   80 m= 0·   km   87 m= 0·   km

L   98 m= 0·   km   100 m= 0·   km   101 m= 0·   km

M   372 m= 0·   km   608 m=   km   716 m= 0·   km

N   930 m=   km   993 m=   km   1 000 m=   km

O   504 m=   km   1 010 m=   km   1 100 m=   km

P   4 608 m=   km   2 072 m=   km   3 806 m=   km

Q   1 040 m=   km   808 m=   km   792 m=   km

R   2 km 321 m=   km   1 km 207 m=   km   3 km 80 m =   km

S   1 km 206 m=   km   2 km 500 m=   km   2 km 30 m =   km

T   3 km 90 m =   km   1 km 809 m=   km   4 km 700 m=   km

U   2 km 890 m=   km   5 km 750 m=   km   1 km 9 m =   km

Kilometres to metres

A   State the value of the seven in each of these numbers:

1 173 m   742 m   2 706 m   7 049 m   8 017 m

B   State how many hundreds there are in   650,   2 118   1 073

C   State how many tens there are in   35,   312,   507

D   State how many tenths there are in   0·71,   1·32,   1·06,

E   State how many hundredths there are in   0·03,   1·046,   0·205

F   State how many thousandths there are in   0·003,   0·215,   0·0301

G   To change kilometres to metres we   by   .

H   To multiply numbers by one thousand we move the figures   places to the   .

Complete:

I	1 km 200 m=	m	2 km 409 m=	m	1 km 270 m=	m
J	1 km 90 m =	m	1 km 86 m =	m	2 km 75 m =	m
K	2 km 300 m=	m	2 km 70 m =	m	4 km 800 m=	m
L	3 km 65 m =	m	1 km 10 m =	m	1 km 18 m =	m
M	1 km 600 m=	m	1 km 705 m=	m	2 km 96 m =	m
N	2 km 89 m =	m	3 km 90 m =	m	1 km 9 m =	m
O	2 km =	m	0·5 km =	m	1·5 km =	m
P	2·345 km =	m	1·078 km =	m	2·086 km =	m
Q	3·897 km =	m	2·809 km =	m	3·904 km =	m
R	1·088 km =	m	2·096 km =	m	4·86 km =	m
S	2·75 km =	m	1·81 km =	m	1·706 km =	m
T	1·7 km =	m	2·3 km =	m	2·9 km =	m
U	0·635 km =	m	0·48 km =	m	0·8 km =	m
V	0·089 km =	m	0·96 km =	m	0·092 km =	m
W	2·08 km =	m	0·76 km =	m	1·05 km =	m

Which of these amounts contain one metre or more?
A   230 cm   89 cm   90 cm   125 cm   650 mm   2750 mm   110 cm

Which of these amounts contain less than one metre?
B   104 cm   140 cm   75 cm   80·6 cm   970 cm   69 cm   9·86 cm

Which of these amounts contain one kilometre or more?
C   1·008 km   560 m   2703 m   6000 m   800 m   2090 m

D   Which number is half-way from 1 to 9?

When writing numbers to the nearest 10 we take all from 5 and over up to the next ten, and discard all below 5.
        i.e. 25 to the nearest ten becomes 30 and 105 becomes 110.

Write to the nearest centimetre:
E   25 mm   126 mm   87 mm   209 mm
     95 mm   198 mm

F   109 mm   80 mm   105 mm   297 mm
     909 mm   97 mm

Write in metres, but to the nearest centimetre:
G   1·168 m   0·928 m   4·273 m   1·296 m   3·596 m

H   0·377 m   0·398 m   0·996 m   0·907 m   0·098 m

I   3·095 m   1·926 m   1·998 m   3·996 m   2·008 m

J   1·198 m   0·905 m   2·096 m   4·097 m   1·809 m

Write these amounts to the nearest kilometre:
K   203·75 km   470·49 km   308·5 km   700·81 km

L   300·604 km   109·396 km   409·706 km   209·906 km

M   99·706 km   199·68 km   509·81 km   909·09 km

Write these amounts in kilometres to the nearest 10 metres:
N   3·644 km   6·704 km   4·087 km   2·096 km

O   1·9096 km   0·198 km   1·909 km   0·997 km

P   1 km 604 m   1 km 708 m   2 km 196 m
     1 km 200 m

Q   2 km 807 m   2 km 98 m   1 km 95 m
     3 km 908 m

R   3 km 996 m   0 km 995 m   2 km 306 m
     1 km 504 m

Check the four rules of length

	Further practice pages
Addition: Write in columns and add: **A** 4 m 16 cm + 10 m 8 cm + 90 cm + 3 m 5 cm **B** 2 km 83 m + 607 m + 3 km 90 m + 1 060 m	34 A–G
Write in columns, expressing all items in the same denomination as that of the largest unit in the row, and add: **C** 20 mm + 3 cm 8 mm + 16 mm + 14 cm 6 mm **D** 872 m + 0·86 km + 2 040 m + 1·8 km	34 H–N
Subtraction: **E** 7 m 9 cm − 5 m 46 cm 2 km 7 m − 98 m	35 A–H
In these pairs subtract the smaller from the larger: **F** 7·4 km and 7 km 45 m 6·35 km and 1 508 m	35 I–P

Multiplication:
G 2 cm 6 mm × 4   5 m 27 cm × 6   4 m 8 cm × 9
  =   cm   =   m   =   m
  × 4   × 6   × 9
	36 A–D

H 1 km 206 m × 7   3 km 85 m × 8   2 km 85 m × 12
  =   km   =   km   =   km
  × 7   × 8   × 12
	36 E–G

Division
I Take answers correct to the 2nd decimal place:
  6 m 24 cm ÷ 5   4 m 33 cm ÷ 6   5 m 8 cm ÷ 9
 	
  = 5)   m   = 6)   m   = 9)   m
	37 A–D

J Take answers correct to the 3rd decimal place:
  4 km 306 m ÷ 7   10 km 86 m ÷ 11   5 km 90 m ÷ 12
 	
  = 7)   km   = 11)   km   = 12)   km
	37 E–H

A

cm	mm
2	6
3	7
1	9

cm	mm
1	1·3
2	3·4
1	6·2

cm	mm
3	2·5
1	6·7
2	4·3

m	cm
4	26
2	40
5	87

m	cm
6	42
	79
2	90

B

m	cm
2	56·5
	48·4
1	70·6

km	m
2	307
5	86
	194

km	m
3	617
	98
2	705

km	m
2	142
	635
4	83

km	m
1	2
3	80
	786

Write in columns and add:

C   3 cm 5 mm+10 cm 6 mm+4 cm 8 mm

D   2 m 38 cm+12 m 75 cm+89 cm+3 m 8 cm

E   1 km 219 m+5 km 408 m+760 m

F   3 km 46 m+708 m+2 km 490 m+1 km 634 m

G   2 km 90 m+30 km 500 m+2 500 m+1 km 8 m

H

cm	m	m	m	km	km	km
3·16	2·8	2·145	6·438	2·142	1·02	2·765
2·7	1·49	0·867	10·5	0·635	30·8	0·089
0·49	3·06	7·55	0·875	4·083	0·786	7·456

Write in columns, expressing all items in the same denomination as that of the largest unit in the row, and add:

I   30 mm+4 cm 6 mm+17 mm+7 cm 5 mm

J   89 cm+127 cm+1·6 m+93 mm

K   30·75 m+0·125 m+40 cm+108·5 cm

L   40·6 m+280 m+2·375 km+0·69 km

M   25·8 km+809 m+6·09 km+95 m

N   870 m+0·86 km+1 090 m+3·07 km

A

cm mm	cm mm	m cm	m cm	m cm
3 2	5 0	4 20	7 0	5 8
−1 8	−2 8	−2 60	−3 76	− 43

B

m cm	km m	km m	km m	km m
7 13	3 60	2 40	4 0	6 0
−4 96	−1 281	− 76	−3 607	−5 87

Subtract:

C 11 cm 4 mm − 2 cm 7 mm   8 cm 3 mm − 6 cm 5 mm

D 4 m 13 cm − 2 m 8 cm   6 m 4 cm − 3 m 37 cm

E 6 m − 4 m 39 cm   5 m 6 cm − 2 m 8 cm

F 8 km 132 m − 5 km 604 m   10 km 63 m − 784 m

G 5 km 104 m − 3 km 697 m   8 km 42 m − 6 km 95 m

H 10 km − 2 km 391 m   2 km 8 m − 99 m

I

mm	cm	m	m	km	km	km
6·3	5·14	6·1	4·2	0·76	2·03	2·06
−3·7	−3·16	−2·15	−0·86	−0·085	−0·935	−1·092

In these pairs subtract the smaller from the larger:

J 2·5 m and 1·2 cm   8·5 cm and 8 cm 6 mm

K 2·46 m and 2·5 m   3·6 m and 3 m 40 cm

L 19 km 60 m and 19·8 km   20·65 km and 20 km 500 m

M 37 m 8 cm and 37·62 m   73 cm 4 mm and 73·04 cm

N 1 km 70 m and 10 700 m   4·06 km and 4·101 km

O 3 km 42 m and 3·401 km   1·1 km and 897 m

P 5 km 89 m and 5·19 km   1·9 km and 291 m

A 2 cm 3 mm ✕ 7   4 cm 2 mm ✕ 6   3 cm 7 mm ✕ 8   5 cm 6 mm ✕ 9
   =   cm   =   cm   =   cm   =   cm
   ✕ 7   ✕ 6   ✕ 8   ✕ 9

B 6 cm 2·5 mm ✕ 5   8 cm 4·7 mm ✕ 9   10 cm 8 mm ✕ 11   10 cm 6·5 mm ✕ 12
   =   cm   =   cm   =   cm   =   cm
   ✕ 5   ✕ 9   ✕ 11   ✕ 12

C 3·14 m ✕ 6   4 m 35 cm ✕ 8   6 m 50 cm ✕ 12   5 m 73 cm ✕ 9
   =   m   =   m   =   m   =   m
   ✕ 6   ✕ 8   ✕ 12   ✕ 9

D 2 m 8 cm ✕ 7   5 m 9 cm ✕ 9   7 m 80 cm ✕ 11   10 m 7 cm ✕ 12
   =   m   =   m   =   m   =   m
   ✕ 7   ✕ 9   ✕ 11   ✕ 12

E 1 km 231 m ✕ 4   1 km 305 m ✕ 6   2 km 410 m ✕ 7   2 km 600 m ✕ 9
   =   km   =   km   =   km   =   km
   ✕ 4   ✕ 6   ✕ 7   ✕ 9

F 2 km 62 m ✕ 5   4 km 73 m ✕ 8   3 km 90 m ✕ 11   5 km 85 m ✕ 12
   =   km   =   km   =   km   =   km
   ✕ 5   ✕ 8   ✕ 11   ✕ 12

G 1 km 4 m ✕ 3   1 km 7 m ✕ 6   2 km 9 m ✕ 12   3 km 9 m ✕ 12
   =   km   =   km   =   km   =   km
   ✕ 3   ✕ 6   ✕ 12   ✕ 12

A 6 cm 5 mm ÷ 5   7 cm 4 mm ÷ 4   6 cm ÷ 8   8 mm ÷ 5

= 5)‾‾‾ cm   = 4)‾‾‾ cm   = 8)‾‾‾ cm   = 5)‾‾‾ cm

Take answers correct to the second decimal place:

B 8 cm 4 mm ÷ 6   5 cm 9 mm ÷ 7   6 cm 8 mm ÷ 11   17 cm 8 mm ÷ 9

= 6)‾‾‾ cm   = 7)‾‾‾ cm   = 11)‾‾‾ cm   = 9)‾‾‾ cm

C 9 m 25 cm ÷ 5   10 m 18 cm ÷ 6   7 m 60 cm ÷ 9   8 m 87 cm ÷ 12

= 5)‾‾‾ m   = 6)‾‾‾ m   = 9)‾‾‾ m   = 12)‾‾‾ m

D 8 m 8 cm ÷ 7   23 m 9 cm ÷ 9   8 m 9 cm ÷ 11   11 m 8 cm ÷ 12

= 7)‾‾‾ m   = 9)‾‾‾ m   = 11)‾‾‾ m   = 12)‾‾‾ m

Take answers correct to the third decimal place:

E 6 m 37 cm ÷ 5   3 m 80 cm ÷ 6   4 m 68 cm ÷ 7   1 m 89 cm ÷ 8

= 5)‾‾‾ m   = 6)‾‾‾ m   = 7)‾‾‾ m   = 8)‾‾‾ m

F 6 km 206 m ÷ 4   10 km 470 m ÷ 7   5 km 817 m ÷ 6   3 km 792 m ÷ 8

= 4)‾‾‾ km   = 7)‾‾‾ km   = 6)‾‾‾ km   = 8)‾‾‾ km

G 6 km 68 m ÷ 5   12 km 70 m ÷ 6   5 km 83 m ÷ 9   4 km 90m ÷ 11

= 5)‾‾‾ km   = 6)‾‾‾ km   = 9)‾‾‾ km   = 11)‾‾‾ km

H 3 km 9 m ÷ 7   2 km 6 m ÷ 9   4 km 4 m ÷ 11   6 km 7 m ÷ 12

= 7)‾‾‾ km   = 9)‾‾‾ km   = 11)‾‾‾ km   = 12)‾‾‾ km

net weight on the outside of a tin, packet, etc., means the weight of the contents of the tin, packet, etc. The total weight of the container and the contents is the **gross weight**.

A   Borrow some packets, tins, etc., which show the net weight and find the gross weight. From that find the weight of the container and the wrappings.

B   Would you say that anything weighing one gramme was heavy, very heavy, light or very light?

The cheese is 32 p per kilo.

C   What is kilo short for?

D   What do you know about a kilo in relation to a gramme?

E   What would ½ kilo of cheese cost?

F   What would 500 grammes of cheese cost?

State which of the articles in this list you would expect to have with the weight shown in kilogrammes (kg) or grammes (g).

G   a bag of coal

H   a tin of baked beans

I   a tin of condensed milk

J   a sack of flour

K   a sack of potatoes

L   your own weight

M   a bunch of seven bananas

N   a packet of custard powder

O   a tablet of soap

	Further practice pages

Complete:

A State the total weight when the weights on a pan are
1 kg, 200 g, 100 g, 50 g, 5 g
2 kg, 500 g, 50 g, 20 g, 2 g   *40 A–L*

B $\frac{1}{2}$ kg =   g $\frac{1}{5}$ kg =   g $\frac{3}{10}$ kg =   g   *41 A–N*

C 500 g =   kg   600 g =   kg   750 g =   kg   *41 O–V*

D 1·326 kg =   g   2·076 kg =   kg   g
0·087 kg =   g   2·005 kg =   kg   g   *42 A–K*

E 1·5 kg =   g   0·77 kg =   kg   g
0·39 kg =   g   1·8 kg =   kg   g   *42 L–U*

F 1 732 g =   kg   2 401 g =   kg   g
206 g =   kg   79 g =   kg   *43 A–K*

G 1 600 g =   kg   2 700 g =   kg   g
2 070 g =   kg   400 g =   kg   *43 L–U*

H

0	I kg	2 kg	3 kg	4 kg	5 kg	6 kg	7 kg	8 kg	9 kg	10 kg

a   b   c   d   e

State in kilogrammes the weights represented on this graduated scale when the pointer is at

a   b   c   d   e   *44 and 45*

I 1 m tonne 246 kg =   kg   2 m tonnes 85 kg =   kg   *46 A–F*

J 1·7 m tonne =   kg   0·08 m tonne =   kg   *46 G–P*

K 604 kg =   m tonne   94 kg =   m tonne   *47 A–I*

L 800 kg =   m tonne   70 kg =   m tonne   *47 J–S*

Using weights

Here are diagrams of most of the weights used on school scales:

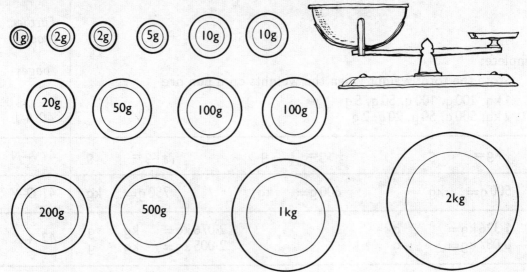

State which weights would be used to weigh

| A | 120g | 70g | 35g |

| B | 300g | 150g | 33g |

| C | 160g | 230g | 90g |

| D | 80g | 75g | 235g |

Only light-weight articles will be weighed in grammes.
Your mother will buy potatoes, meat, fruit, etc., in kilogrammes.

State the weights which would be used to weigh:

E $\frac{1}{2}$ kg $\quad\quad\quad\quad\quad\quad$ $2\frac{1}{2}$ kg
$\frac{1}{4}$ kg

F $\frac{3}{4}$ kg $\quad\quad\quad\quad\quad\quad$ $3\frac{1}{2}$ kg
$2\frac{1}{4}$ kg

G 1 kg 40 g $\quad\quad\quad\quad\quad$ 2 kg 80 g
950 g

H 850 g $\quad\quad\quad\quad\quad\quad$ 1 kg 60 g
2 kg 125 g

State the weight of a parcel when the weights on the pan are

I $\quad$ 100 g, 50 g, 20 g, 2 g $\quad\quad\quad$ 200 g, 100 g, 100 g, 20 g, 10 g

J $\quad$ 500 g, 200 g, 100 g, 20 g, 5 g $\quad\quad$ 1 kg, 50 g, 20 g, 10 g, 10 g, 2 g

K $\quad$ 200 g, 100 g, 100 g, 50 g, 20 g $\quad\quad$ 2 kg, 500 g, 50 g, 10 g, 5 g, 2 g, 1 g

L $\quad$ 500 g, 100 g, 50 g, 20 g, 5 g, 2 g $\quad\quad$ 2 kg, 1 kg, 500 g, 20 g, 10 g, 2 g

Complete. Work across the page:

A   1 kilogramme=   grammes $\frac{1}{2}$ kilogramme=   grammes

B $\frac{1}{4}$ kilogramme=   grammes $\frac{3}{4}$ kilogramme=   grammes

C $\frac{1}{10}$ kilogramme=   grammes $\frac{1}{5}$ kilogramme=   grammes

D   500 grammes=   kilogramme   100 grammes=   kilogramme

E   200 grammes=   kilogramme   250 grammes=   kilogramme

F $\frac{1}{4}$ kg=   g $\frac{3}{4}$ kg=   g $\frac{1}{2}$ kg=   g $\frac{1}{5}$ kg=   g

G $\frac{2}{5}$ kg=   g $\frac{1}{10}$ kg=   g $\frac{3}{4}$ kg=   g $\frac{1}{4}$ kg=   g

H $1\frac{1}{2}$ kg=   g $1\frac{1}{4}$ kg=   g $\frac{2}{5}$ kg=   g $\frac{3}{5}$ kg=   g

I $\frac{1}{10}$ kg=   g $\frac{3}{10}$ kg=   g $\frac{7}{10}$ kg=   g $\frac{3}{4}$ kg=   g

J $\frac{1}{4}$ kg=   g $\frac{2}{5}$ kg=   g $\frac{4}{5}$ kg=   g $\frac{3}{10}$ kg=   g

K $\frac{7}{10}$ kg=   g $\frac{9}{10}$ kg=   g $\frac{3}{5}$ kg=   g $\frac{4}{5}$ kg=   g

---

L $\frac{1}{5}$ kg= $\frac{}{10}$ kg=   g $\frac{4}{5}$ kg= $\frac{}{10}$ kg=   g

M $\frac{3}{5}$ kg= $\frac{}{10}$ kg=   g $\frac{2}{5}$ kg= $\frac{}{10}$ kg=   g

N $\frac{4}{5}$ kg= $\frac{}{10}$ kg=   g $\frac{3}{5}$ kg= $\frac{}{10}$ kg=   g

Express your answers first as a fraction and then as a decimal:

O   200 g=   kg   or   kg   400 g=   kg   or   kg

P   800 g=   kg   or   kg   600 g=   kg   or   kg

Q   400 g=   kg   or   kg   800 g=   kg   or   kg

R   600 g=   kg   or   kg   200 g=   kg   or   kg

Express answers as decimals:

S   300 g=   kg   500 g=   kg   70 g=   kg   100 g=   kg

T   700 g=   kg   200 g=   kg   30 g=   kg   500 g=   kg

U   550 g=   kg   350 g=   kg   150 g=   kg   650 g=   kg

V   800 g=   kg   850 g=   kg   250 g=   kg   750 g=   kg

Complete:

A   To change kilogrammes to grammes we   by   .

B   To multiply a number by one thousand we move the figures   places to the
.

C   1·215 kilogrammes=   grammes   1·706 kilogrammes=   grammes

D   2·307 kilogrammes=   grammes   3·462 kilogrammes=   grammes

E   0·422 kg=   g   0·301 kg=   g   0·702 kg=   g

F   0·291 kg=   g   0·087 kg=   g   0·012 kg=   g

G   0·504 kg=   g   0·062 kg=   g   0·035 kg=   g

H   0·072 kg=   g   0·007 kg=   g   0·006 kg=   g

I   1·083 kg=   g   0·008 kg=   g   0·009 kg=   g

J   1·326 kg=   kg   g   2·081 kg=   kg   g

K   2·005 kg=   kg   g   1·007 kg=   kg   g

L   1·621 kg=   g   1·62 kg=   g   1·34 kg=   g

M   1·73 kg =   g   2·84 kg=   g   0·76 kg=   g

N   0·69 kg =   g   0·16 kg=   g   0·37 kg=   g

O   0·41 kg =   g   0·28 kg=   g   0·85 kg=   g

P   0·77 kg =   g   0·7 kg =   g   0·5 kg =   g

Q   0·832 kg=   g   0·8 kg =   g   0·6 kg =   g

R   1·5 kg=   kg   g   2·9 kg=   kg   g

S   3·4 kg=   kg   g   1·7 kg=   kg   g

Which is the heavier in each of these pairs?

T $\frac{1}{2}$ kg   or   51 g $\frac{1}{4}$ kg   or   202 g $\frac{1}{2}$ kg   or   650 g

U $\frac{3}{4}$ kg   or   750 g $\frac{3}{4}$ kg   or   75 g $\frac{1}{4}$ kg   or   226 g

V   0·76 kg or 700 g   1·4 kg or 1 420 g   0·85 kg or 800 g

W   0·076 kg or 175 g   0·806 kg or 860 g   0·27 kg or 30 g

Complete:

A To multiply a number by one thousand we move the figures ⎯⎯ places to the ⎯⎯ .

B To divide a number by one thousand we move the figures ⎯⎯ places to the ⎯⎯ .

C To change grammes into kilogrammes we ⎯⎯ by moving the figures ⎯⎯ places to the ⎯⎯ .

D 2 426 grammes = ⎯⎯ kilogrammes ⎯⎯ grammes   345 g = ⎯⎯ kg ⎯⎯ g

E 1 095 grammes = ⎯⎯ kilogrammes ⎯⎯ grammes   207 g = ⎯⎯ kg ⎯⎯ g

F 2 089 grammes = ⎯⎯ kilogrammes   1 276 grammes = ⎯⎯ kilogrammes

G	1 732 g = ⎯ kg	732 g = ⎯ kg	845 g = ⎯ kg
H	1 206 g = ⎯ kg	206 g = ⎯ kg	408 g = ⎯ kg
I	2 083 g = ⎯ kg	83 g = ⎯ kg	76 g = ⎯ kg
J	1 071 g = ⎯ kg	71 g = ⎯ kg	69 g = ⎯ kg
K	58 g = ⎯ kg	92 g = ⎯ kg	83 g = ⎯ kg
L	1 006 g = ⎯ kg	6 g = ⎯ kg	8 g = ⎯ kg
M	1 009 g = ⎯ kg	7 g = ⎯ kg	5 g = ⎯ kg
N	2 450 g = ⎯ kg	620 g = ⎯ kg	490 g = ⎯ kg
O	1 670 g = ⎯ kg	670 g = ⎯ kg	510 g = ⎯ kg
P	2 080 g = ⎯ kg	80 g = ⎯ kg	40 g = ⎯ kg
Q	70 g = ⎯ kg	60 g = ⎯ kg	90 g = ⎯ kg
R	1 500 g = ⎯ kg	500 g = ⎯ kg	800 g = ⎯ kg
S	400 g = ⎯ kg	900 g = ⎯ kg	300 g = ⎯ kg

Which is the lighter in each of these pairs?

T $1\frac{1}{2}$ kg or 1 250 g $\frac{3}{4}$ kg or 800 g $\frac{1}{4}$ kg or 205 g

U $1\frac{1}{4}$ kg or 1 500 g $1\frac{1}{2}$ kg or 1 500 g $\frac{1}{2}$ kg or 750 g

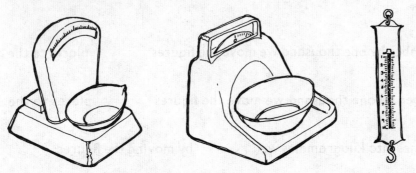

Usually your school scales work by balancing the article to be weighed against marked weights made of brass or iron.

Weighing in shops and industry is usually done on machines which make use of levers or springs or both to move a pointer along a marked scale or to turn a marked disc past a special mark on the machine.

Here is a simple scale:

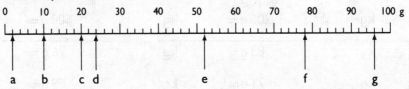

A   What is the heaviest weight that can be shown on this scale?

B   What weight is shown by the pointer at b?

C   What weight does each of the smallest divisions show?

D   State the weight of an article when the pointer is at
a   c   d   e   f   g

Look carefully at this scale:

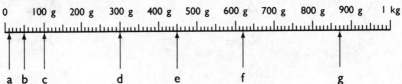

E   What is the heaviest weight that can be shown on this scale?

F   What weight is shown by the pointer at c?

G   What weight does each of the smallest divisions show?

H   State the weight of an article when the pointer is at
a   b   d   e   f   g

More graduated scales

Look carefully at the scales on this page, then answer the questions.

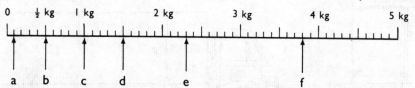

A What is the heaviest weight that can be shown on this scale?

B What is the weight shown at c?

C What is the weight represented by each of the smallest divisions in grammes? in kilogrammes?

D State in kilogrammes, first as a fraction and then as a decimal, the weights shown when the pointer is at

a b d e f

E What is the heaviest weight that can be shown on this scale?

F What is the weight shown at c?

G What is the weight shown at a? at b?

H State in kilogrammes, first as a fraction and then as a decimal, the weights shown when the pointer is at

d e f g h

I What weight is indicated by the pointer at d? at a?

J What weight is indicated by one of the smallest divisions? Give your answer first as a fraction and then as a decimal.

K State in kilogrammes the weight indicated by the pointer at

b c e f g h

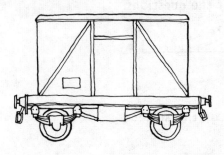 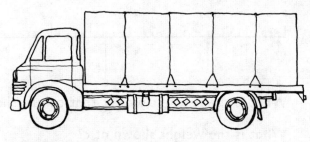

1 metric tonne = 1 000 kilogrammes

Complete:

A To change metric tonnes into kilogrammes we move the figures ____ places to the ____ .

B 1 tonne 341 kg = ____ kg   2 tonne 456 kg = ____ kg

C 1 tonne 750 kg = ____ kg   1 tonne 801 kg = ____ kg

D 2 tonne 870 kg = ____ kg   1 tonne 800 kg = ____ kg

E 2 tonne 96 kg = ____ kg   1 tonne 78 kg = ____ kg

F 1 tonne 80 kg = ____ kg   1 tonne 90 kg = ____ kg

G 1·327 tonne = 1 tonne ____ kg = ____ kg   1·516 tonne = ____ kg

H 0·476 tonne = ____ tonne ____ = ____ kg   0·923 tonne = ____ kg

I 0·083 tonne = ____ kg   0·076 tonne = ____ kg   0·028 tonne = ____ kg

J 0·307 tonne = ____ kg   0·705 tonne = ____ kg   0·064 tonne = ____ kg

K 1·63 tonne = ____ tonne ____ kg = ____ kg   1·72 tonne = ____ kg

L 1·91 tonne = ____ tonne ____ kg = ____ kg   1·58 tonne = ____ kg

M 0·76 tonne = ____ kg   0·38 tonne = ____ kg   0·65 tonne = ____ kg

N 1·065 tonne = ____ kg   0·072 tonne = ____ kg   0·096 tonne = ____ kg

O 1·5 tonne = ____ kg   0·7 tonne = ____ kg   0·8 tonne = ____ kg

P 1·008 tonne = ____ kg   0·08 tonne = ____ kg   0·06 tonne = ____ kg

A To change kilogrammes into metric tonnes we move the figures ___ places to the ___ .

B 1 264 kg = ___ tonne ___ kg = ___ tonnes   1 376 kg = ___ tonnes

C 1 378 kg = ___ tonne ___ kg = ___ tonnes   2 649 kg = ___ tonnes

D 426 kg = ___ tonne   723 kg = ___ tonne   487 kg = ___ tonne

E 407 kg = ___ tonne   480 kg = ___ tonne   856 kg = ___ tonne

---

F 1 147 kg = ___ tonne ___ kg = ___ tonnes   208 kg = ___ tonne

G 1 086 kg = ___ tonne ___ kg = ___ tonnes   1 067 kg = ___ tonnes

H 73 kg = ___ tonne   82 kg = ___ tonne   96 kg = ___ tonne

I 65 kg = ___ tonne   91 kg = ___ tonne   78 kg = ___ tonne

J 1 060 kg = ___ tonne ___ kg = ___ tonnes   50 kg = ___ tonne

K 2 010 kg = ___ tonnes ___ kg = ___ tonnes   80 kg = ___ tonne

L 1 400 kg = ___ tonne ___ kg = ___ tonnes   1 100 kg = ___ tonnes

M 500 kg = ___ tonne   200 kg = ___ tonne   800 kg = ___ tonne

N 900 kg = ___ tonne   90 kg = ___ tonne   60 kg = ___ tonne

O 600 kg = ___ tonne   809 kg = ___ tonne   400 kg = ___ tonne

P 1 200 kg = ___ tonnes   2 760 kg = ___ tonnes   1 080 kg = ___ tonnes

---

If these weights were on the pan of your scales state the total each time in kilogrammes:

Q 1 kg, 50 g, 100 g, 100 g   2 kg, 500 g, 50 g, 20 g, 10 g

R 500 g, 100 g, 50 g, 5 g, 2 g   1 kg, 500 g, 200 g, 5 g, 1 g

S 200 g, 100 g, 100 g, 2 g, 2 g, 1 g   50 g, 20 g, 10 g, 10 g

		Further practice pages

Addition:

A

kg	g	tonnes	kg	kg	tonnes
3	207	4	850	0·376	0·72
1	98		706	2·78	2·8
	803	1	98	0·059	0·093

49 A–F

B Write in columns, expressing all items in the same denomination as that of the largest unit, and add:

3 kg 75 g + 1 kg 827 g + 750 g + 2½ kg
2 kg 700 g + 88 g + 1·06 kg + 90 g
3 tonnes 850 kg + 1 050 kg + 1 tonne 75 kg + 2¼ tonnes

49 G–L

Subtraction:

C

kg	g	tonnes	kg	kg	tonnes
2	106	3	46	2·06	4·05
−1	508	−	398	−0·566	−2·075

50 A–C

D In each pair take the smaller from the larger:

1·05 tonne and 150 kg   3 tonnes and 2 090 kg
0·65 kg and 587 grammes   ½ kg and 55 grammes

50 D–Q

Multiplication:

E

kg	g		kg	g		tonnes	kg
2	600×8		1	78×9		3	95×12
=	kg		=	kg		=	tonnes
×	8		×	9		×	12

51 A–G

Division:

F Take answers to the third decimal place:

2 kg 800 g ÷ 6   1 kg 75 g ÷ 9   3 tonnes 450 kg ÷ 11
=   =   =
6)⎺⎺⎺⎺ kg   9)⎺⎺⎺⎺ kg   11)⎺⎺⎺⎺ tonnes

52 A–C

G Take answers correct to the third decimal place:

4 kg 75 g ÷ 7   1 kg 86 g ÷ 11   3 tonnes 790 kg ÷ 12
=   =   =
7)⎺⎺⎺⎺ kg   11)⎺⎺⎺⎺ kg   12)⎺⎺⎺⎺ tonnes

52 D–H

A

kg	g		kg	g		kg	g		kg	g		kg	g
1	245		2	473		1	603		1	93		3	406
2	136			809		3	98			450		1	500
1	675		3	52			106		5	808			98

B

tonnes	kg		tonnes	kg		tonnes	kg		tonnes	kg		tonnes	kg
4	96			700		4	83		2	50		1	76
	704		5	98		1	9		1	85		5	58
1	800		2	606			90			98			9

Write in columns and add:

C 2 tonnes 70 kg + 690 kg + 1 tonne 806 kg + 78 kg

D 3 kg 500 g + 896 g + 738 g + 1 kg 85 g

E 750 g + 4 kg 860 g + 75 g + 98 g

F

kg	kg	kg	kg	tonnes	tonnes
1·623	2·085	1·6	0·75	0·07	0·8
0·081	0·726	0·508	0·08	2·697	0·096
3·76	0·09	0·096	0·97	0·045	0·79

Write in columns, expressing all items in the same denomination as that of the largest unit in the row, and add:

G 6 kg 706 g + 2 kg 68 g + 409 g + $1\frac{1}{2}$ kg

H 2 kg 800 g + 90 g + 1 kg 700 g + $2\frac{1}{4}$ kg

I 3 tonnes + 1 400 kg + 876 kg + 1 085 kg

J 2 tonnes 76 kg + 3 tonnes 894 kg + 85 kg + 690 kg

K 608 kg + 96 kg + 4 tonnes 700 kg + 2 088 kg

L 760 g + 1 085 g + 2 kg 200 g + 95 g + 1 080 g

A

kg g	kg g	tonnes kg	tonnes kg	tonnes	kg
5 607	4 700	7 650	3 85	2·46	1·62
−2 388	−3 806	−4 842	− 778	−1·075	−0·706

B

kg g	kg g	tonnes kg	tonnes kg	tonnes	kg
4 80	1 65	1 40	5 200	1·62	4
− 792	− 95	− 794	−2 995	−0·843	−1·641

C

tonnes	tonnes	tonnes	kg	kg	kg
2·675	1·076	3·006	4·008	3·204	5·07
−0·685	−0·808	−1·892	−0·891	−1·906	−0·995

In each of these pairs subtract the smaller item from the larger:

D   1·35 tonnes and 1 tonne 35 kilogrammes

E   1·7 tonnes and 700 kilogrammes

F   2·07 kilogrammes and 2 kilogrammes 700 grammes

G   1·33 kilogrammes and 1 400 grammes

H   2·5 tonnes and 2 tonnes 495 kg   2·1 tonnes and 296 kg

I   0·56 tonnes and 601 kg   1·086 kg and 976 g

J   1·078 kg and 1·102 kg   2·706 kg and 2·74 kg

K   0·604 tonnes and 0·64 tonnes   0·096 kg and 0·101 kg

L   0·701 kg and 698 grammes   1·2 tonnes and 1 196 kg

M   4·2 tonnes and 976 kg   0·063 tonnes and 480 kg

N   1·006 tonnes and 850 kg   2 tonnes and 19 205 kg

O   708 grammes and 7·08 kilos   3 kilos and 30 000 grammes

P $\frac{1}{2}$ kilo and 50 grammes $1\frac{1}{2}$ kilos and 1 500 grammes

Q $\frac{1}{5}$ tonne and 500 kilos $\frac{2}{5}$ tonne and 250 kilos

Multiplication

A

kg	g	
2	132×6	

= _____ kg
× _____ 6

kg	g	
1	308×7	

= _____ kg
× _____ 7

kg	g	
3	96×4	

= _____ kg
× _____ 4

kg	g	
2	87×9	

= _____ kg
× _____ 9

B

tonnes	kg
2	609×8

= _____ tonnes
× _____ 8

tonnes	kg
1	89×9

= _____ tonnes
× _____ 9

tonnes	kg
3	709×12

= _____ tonnes
× _____ 12

tonnes	kg
2	98×11

= _____ tonnes
× _____ 11

C

0·806 kg
× _____ 7

0·076 tonnes
× _____ 9

1·094 tonnes
× _____ 12

1·506 kg
× _____ 8

2·009 kg
× _____ 12

D

kg	g	
1	206×5	

= _____ kg
× _____ 5

kg	g	
0	768×7	

= _____ kg
× _____ 7

kg	g	
0	95×8	

= _____ kg
× _____ 8

kg	g	
0	89×11	

= _____ kg
× _____ 11

E

kg	g	
1	78×7	

= _____ kg
× _____ 7

kg	g	
2	69×12	

= _____ kg
× _____ 12

kg	g	
1	80×11	

= _____ kg
× _____ 11

kg	g	
3	400×12	

= _____ kg
× _____ 12

F

tonnes	kg
1	807×5

= _____ tonnes
× _____ 5

tonnes	kg
2	80×5

= _____ tonnes
× _____ 5

tonnes	kg
1	700×9

= _____ tonnes
× _____ 9

tonnes	kg
2	806×12

= _____ tonnes
× _____ 12

G

tonnes	kg
3	708×7

= _____ tonnes
× _____ 7

tonnes	kg
1	95×9

= _____ tonnes
× _____ 9

tonnes	kg
2	895×12

= _____ tonnes
× _____ 12

tonnes	kg
3	708×11

= _____ tonnes
× _____ 11

52 Division

Take answers to the third decimal place:

A   5 kg 236 g ÷ 4   7 kg 406 g ÷ 6   9 kg 500 g ÷ 7   2 kg 805 g ÷ 8

     = 4)   kg   = 6)   kg   = 7)   kg   = 8)   kg

B   3 kg 400 g ÷ 6   2 kg 560 g ÷ 8   1 kg 705 g ÷ 7   3 kg 806 g ÷ 9

     = 6)   kg   = 8)   kg   = 7)   kg   = 9)   kg

C   1 kg 85 g ÷ 4   1 kg 90 g ÷ 6   2 kg 84 g ÷ 8   1 kg 92 g ÷ 7

     = 4)   kg   = 6)   kg   = 8)   kg   = 7)   kg

Take answers correct to the third decimal place:

D   2 kg 175 g ÷ 6   2 kg 800 g ÷ 7   1 kg 604 g ÷ 5   3 kg 700 g ÷ 9

     = 6)   kg   = 7)   kg   = 5)   kg   = 9)   kg

E   1 kg 75 g ÷ 4   1 kg 60 g ÷ 6   2 kg 88 g ÷ 11   1 kg 90 g ÷ 12

     = 4)   kg   = 6)   kg   = 11)   kg   = 12)   kg

F   2 kg 50 g ÷ 7   0 kg 170 g ÷ 4   0 kg 801 g ÷ 9   1 kg 8 g ÷ 6

     = 7)   kg   = 4)   kg   = 9)   kg   = 6)   kg

G   3 tonnes 805 kg ÷ 6   2 tonnes 300 kg ÷ 7   1 tonne 89 kg ÷ 11   1 tonne 100 kg ÷ 12

     = 6)   tonnes   = 7)   tonnes   = 11)   tonnes   = 12)   tonnes

H   2 tonnes 94 kg ÷ 8   3 tonnes 580 kg ÷ 9   1 tonne 869 kg ÷ 11   1 tonne 78 kg ÷ 12

     = 8)   tonnes   = 9)   tonnes   = 11)   tonnes   = 12)   tonnes

Use some thick paper or very thin card to make this model:

A   Draw this shape:

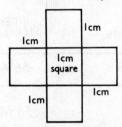

B   Cut out the shape and fix it with Sellotape, folding thus:

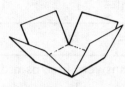

You have a measure having a volume or capacity of one cubic centimetre.
That is also a measure for one **milli-litre.**

C   If you can, obtain some silver sand and find the capacity of
        a teaspoon   a dessert spoon   an egg-cup or small jar.

If a doctor tells you to take medicine in doses of a teaspoonful the chemist will supply
with your bottle a plastic spoon, which measures a dose of 5 millilitres.

D   Use water to find how many school milk bottles are equal to the capacity of the
        home milk bottle.

E   Make another container, cuboid in shape as before, but having a base and sides
        of 10 centimetres square. This has the volume or capacity equal to that of 1 litre.

Put l or ml to show which of these is likely to be measured in litres and which in millilitres:

        F   a bottle of tomato sauce

        G   a churn of milk

        H   a can of petrol

        I   a bottle of eye drops

        J   an ordinary tea-cup

        K   a household bucket

        L   a decorator's paint tin

			Further practice
			pages
A	Complete: 3 000 ml= l 1 070 ml= l 800 ml= l		55 A–V
B	$\frac{1}{2}$ litre= ml $1\frac{1}{4}$ l= ml $\frac{2}{5}$ litre= ml		56 A–L
C	State answers first as a decimal and then as a fraction: 600 ml= l or l 250 ml= l or l		56 M–R
D	Which is the larger in each of these pairs? 0·28 l or 28 ml 1·02 l or 1 200 ml 7·06 l or 760 ml		57 A–L
E	0·75 l= ml 1·6 l= ml 0·08 l= ml		57 M–V
F	What is represented by one division in each of these scales? ─500 ml ─200 ml ─400 ml ─300 ml ─100 ml		58 A–L
G	Write in columns, expressing all items in the same denomination as that of the largest item, and add: 1 l 78 ml+705 ml+86 ml 2 l 700 ml+96 ml+807 ml 3 l 600 ml+1 l 85 ml+2 700 ml+1 l 608 ml		59 A–G
H	In these pairs subtract the smaller from the larger: 1 l 88 ml and 1·08 l 1·2 litres and 120 ml 2 l 70 ml and 2·7 l $\frac{1}{2}$ litre and 550 ml		59 H–K
I	In these pairs subtract the smaller from the larger: 2 l and 200 ml 5 l and 570 ml 0·06 l and 600 ml		60 A–D
J	l ml l ml l ml 2 207×9 4 95×11 3 109×12 = l = l l × 9 × 11 × 12		60 E–I
K	Take answers correct to the third decimal place: 6 l 185 ml÷6 5 l 700 ml÷9 2 l 80 ml÷12 =6) l =9) l =12) l		61 A–H

Complete. Work across the page:

A   1 litre =   millilitres $\frac{1}{2}$ litre =   millilitres

B   500 ml =   litre   1 000 ml =   litre

C   1 litre =   ml   2 litres =   ml   3 litres =   ml

D   2 000 ml =   litres   3 000 ml =   litres   1 000 ml =   litre

---

E   To change millilitres to litres we   by   .

F   To divide numbers by one thousand we move the figures   places to the
        .

G   How many hundreds in one thousand?

H   What part of one thousand is 100 as a fraction?

I   What part of one thousand is 100 as a decimal?

J   What part of one thousand is 200 as a decimal?

K   What part of one thousand is 600 as a decimal?

Divide each of these numbers by 1 000:

L   2 036   2 040   789   650   86   704

M   400   30   90   9   8   70

Complete:

N   2 000 ml =   litres   2 641 ml =   l   ml =   l

O   4 000 ml =   litres   4 185 ml =   l   ml =   l

P   3 000 ml =   litres   3 760 ml =   l   ml =   l

Q   2 500 ml =   litres   2 048 ml =   l   ml =   l

R   1 500 ml =   litres   1 030 ml =   l   ml =   l

S   671 ml =   l   408 ml =   l   860 ml =   l

T   540 ml =   l   93 ml =   l   75 ml =   l

U   100 ml =   l   30 ml =   l   90 ml =   l

V   9 ml =   l   7 ml =   l   80 ml =   l

Complete. Work across the page:

A   1 litre=   millilitres $\frac{1}{2}$ litre=   millilitres

B $\frac{1}{4}$ litre=   millilitres $\frac{3}{4}$ litre=   millilitres

C $\frac{1}{2}$ l=   ml   or   0·   l $\frac{1}{10}$ l=   ml   or   0·   l

D $\frac{1}{4}$ l=   ml   or   l $\frac{3}{4}$ l=   ml   or   l

E $\frac{1}{5}$ l=   ml   or   l $1\frac{1}{4}$ l=   ml   or   l

F $\frac{3}{10}$ l=   ml   or   l $\frac{3}{4}$ l=   ml   or   l

G $1\frac{1}{2}$ l=   ml   or   l $2\frac{2}{10}$ l=   ml   or   l

H $\frac{7}{10}$ l=   ml   or   l $\frac{9}{10}$ l=   ml   or   l

I $2\frac{1}{5}$ l=   ml   or   l $\frac{2}{5}$ l=   ml   or   l

J $\frac{9}{10}$ l=   ml   or   l $\frac{3}{5}$ l=   ml   or   l

K $1\frac{1}{4}$ l=   ml   or   l $1\frac{3}{4}$ l=   ml   or   l

L $3\frac{2}{5}$ l=   ml   or   l $\frac{4}{5}$ l=   ml   or   l

---

State your answers first as a decimal and then as a fraction:

M   500 ml=   l   or   l   200 ml=   l   or   l

N   250 ml=   l   or   l   100 ml=   l   or   l

O   400 ml=   l   or   l   750 ml=   l   or   l

P   600 ml=   l   or   l   700 ml=   l   or   l

Q   900 ml=   l   or   l   300 ml=   l   or   l

R   750 ml=   l   or   l   250 ml=   l   or   l

Complete. State each answer as a decimal:

A	1 000 ml =	l	100 ml =	l	150 ml =	l
B	250 ml =	l	275 ml =	l	307 ml =	l
C	500 ml =	l	505 ml =	l	606 ml =	l
D	100 ml =	l	170 ml =	l	70 ml =	l
E	80 ml =	l	65 ml =	l	55 ml =	l
F	830 ml =	l	601 ml =	l	710 ml =	l
G	400 ml =	l	40 ml =	l	4 ml =	l
H	60 ml =	l	6 ml =	l	9 ml =	l

Which is the smaller in each of these pairs?

I	0·37 l or 37 ml	1·55 l or 1 055 ml	0·8 l or 80 ml
J	0·07 l or 700 ml	1·02 l or 1 200 ml	0·72 l or 700 ml
K	1·27 l or 128 ml	0·706 l or 760 ml	0·08 l or 80 ml
L	3·01 l or 3 100 ml	0·024 l or 240 ml	0·59 l or 590 ml

Complete:

M	0·5 l =	ml	0·271 l =	ml	0·306 l =	ml
N	0·425 l =	ml	0·203 l =	ml	0·405 l =	ml
O	0·046 l =	ml	0·052 l =	ml	0·087 l =	ml
P	0·005 l =	ml	0·006 l =	ml	0·008 l =	ml
Q	0·007 l =	ml	0·07 l =	ml	0·09 l =	ml
R	0·04 l =	ml	1·03 l =	ml	2·08 l =	ml
S	0·5 l =	ml	0·25 l =	ml	1·75 l =	ml
T	0·4 l =	ml	0·7 l =	ml	2·6 l =	ml
U	0·8 l =	ml	4·2 l =	ml	0·9 l =	ml
V	0·203 l =	ml	0·23 l =	ml	2·3 l =	ml

An average teaspoon holds 5 ml.

A bottle of Coca-cola contains approximately 250 ml.

Here is one type of measure you may use.

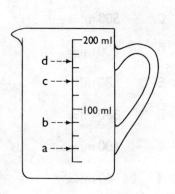

A How many small sections are there between 200 ml and 100 ml?

B What does each section represent?

C How many divisions are there below 100 ml?

D Why?

E What does the lowest mark represent?

F State how much liquid would be in the measure when it reaches the level shown by

    a   b   c   d

Here is another type likely to be found in a kitchen.

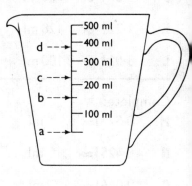

G Think why these divisions are not of the same size and why the widest are nearest the bottom of the measure.

H What does each division represent?

I What does the lowest mark represent?

J State how much liquid would be in the measure when it reaches the level shown by

    a   b   c   d

One of the most common liquids measured is petrol.

K If a car uses 5 litres of petrol to travel 50 kilometres, how much will it use to travel 10 km?

L If a car uses 1·1 litres of petrol per 10 km how much petrol will it need to travel 90 km?

A

l	ml	l	ml	l	ml	l	ml	l	ml
1	324	2	504	1	608	2	94		76
2	86		93	2	75	4	86	2	95
1	705	1	600		106	1	70		58

Write in columns and add:

B 2 l 73 ml + 48 ml + 1 l 700 ml   1 l 49 ml + 96 ml + 48 ml

C 3 l 50 ml + 783 ml + 68 ml   85 ml + 2 l 50 ml + 76 ml

D

litres	litres	litres	litres	litres	litres
7·236	1·835	0·762	1·41	2·76	4·385
0·508	0·092	0·049	0·7	0·8	0·096
0·072	2·408	0·203	0·083	1·69	3·8

Write in columns, expressing all items in the same denomination as that of the largest item in the row, and add:

E 4 l 78 ml + 3 l 806 ml + 1 l 450 ml + 5 l 800 ml

F 3 700 ml + 2 l 90 ml + 460 ml   5 l 750 ml + 80 ml + 900 ml

G 1 l 80 ml + 420 ml + 1 070 ml   2 500 ml + 3 l 670 ml + 75 ml

Subtraction:

H

l	ml	l	ml	l	ml	l	ml	l	ml
3	76	5	800	2	70	3	10	2	0
−1	208	−2	92	−	885	−1	182	−	45

In these pairs subtract the smaller from the larger:

I 4 litres 85 millilitres and 10 700 millilitres

J 3 litres 40 ml and 1 l 242 ml   2 l 180 ml and 4 l 86 ml

K 1 l 175 ml and 4 l 50 ml   3 l 75 ml and 1 l 896 ml

A	litres	litres	litres	litres	litres	litres
	6·7	4·05	2·075	3·026	4	6
	−2·341	−2·075	−0·095	−1·938	−1·8	−3·605

In these pairs subtract the smaller from the larger:

B   3·4 litres and 3 litres 275 millilitres   2 litres and 10 400 ml

C   0·62 litres and 595 millilitres   4 litres and 496 ml

D   1·04 litres and 1 400 millilitres   3 litres and 1 050 ml

Multiplication:

E	l ml	l ml	l ml	l ml
	3 406×6	2 370×8	4 978×11	1 809×12
	= l	= l	= l	= l
	× 6	× 8	× 11	× 12

F	l ml	l ml	l ml	l ml
	2 75×7	1 86×9	3 95×11	2 90×12
	= l	= l	= l	= l
	× 7	× 9	× 11	× 12

G	l ml	l ml	l ml	l ml
	1 600×6	2 80×5	1 70×9	3 80×12
	= l	= l	= l	= l
	× 6	× 5	× 9	× 12

H	l ml	l ml	l ml	l ml
	3 804×7	1 90×8	2 96×11	2 708×9
	= l	= l	= l	= l
	× 7	× 8	× 11	× 9

I	0·628 l	1·008 l	2·09 l	0·083 l	0·092 l
	× 9	× 11	× 12	× 9	× 12

A 3 l 205 ml ÷ 5   2 l 744 ml ÷ 7   10 l 96 ml ÷ 8   14 l 64 ml ÷ 6

= 5)‾‾‾‾‾ l   = 7)‾‾‾‾‾ l   = 8)‾‾‾‾‾ l   = 6)‾‾‾‾‾ l

B 4 l 68 ml ÷ 6   6 l 28 ml ÷ 11   2 l 7 ml ÷ 9   10 l 80 ml ÷ 12

= 6)‾‾‾‾‾ l   = 11)‾‾‾‾‾ l   = 9)‾‾‾‾‾ l   = 12)‾‾‾‾‾ l

Take answers to the third decimal place:
C 7 l 803 ml ÷ 5   6 l 207 ml ÷ 6   4 l 900 ml ÷ 8   5 l 650 ml ÷ 11

= 5)‾‾‾‾‾ l   = 6)‾‾‾‾‾ l   = 8)‾‾‾‾‾ l   = 11)‾‾‾‾‾ l

D 1 l 706 ml ÷ 6   4 l 291 ml ÷ 7   2 l 408 ml ÷ 9   3 l 206 ml ÷ 8

= 6)‾‾‾‾‾ l   = 7)‾‾‾‾‾ l   = 9)‾‾‾‾‾ l   = 8)‾‾‾‾‾ l

E 9 l 80 ml ÷ 7   3 l 49 ml ÷ 6   4 l 97 ml ÷ 11   7 800 ml ÷ 12

= 7)‾‾‾‾‾ l   = 6)‾‾‾‾‾ l   = 11)‾‾‾‾‾ l   = 12)‾‾‾‾‾ l

Take answers correct to the third decimal place:
F 1 l 208 ml ÷ 5   2 l 706 ml ÷ 7   9 l 700 ml ÷ 6   5 l 500 ml ÷ 8

= 5)‾‾‾‾‾ l   = 7)‾‾‾‾‾ l   = 6)‾‾‾‾‾ l   = 8)‾‾‾‾‾ l

G 5 l 80 ml ÷ 6   3 l 90 ml ÷ 8   0 l 807 ml ÷ 11   0 l 98 ml ÷ 12

= 6)‾‾‾‾‾ l   = 8)‾‾‾‾‾ l   = 11)‾‾‾‾‾ l   = 12)‾‾‾‾‾ l

H 7 l 80 ml ÷ 9   5 l 70 ml ÷ 12   3 l 800 ml ÷ 11   9 l 60 ml ÷ 12

= 9)‾‾‾‾‾ l   = 12)‾‾‾‾‾ l   = 11)‾‾‾‾‾ l   = 12)‾‾‾‾‾ l

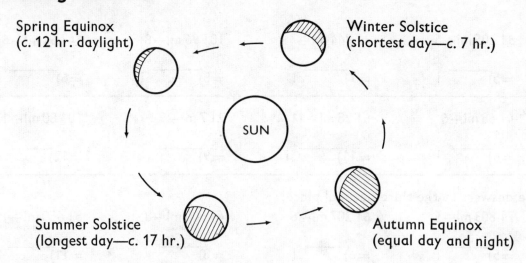

Spring Equinox
(c. 12 hr. daylight)

Winter Solstice
(shortest day—c. 7 hr.)

SUN

Summer Solstice
(longest day—c. 17 hr.)

Autumn Equinox
(equal day and night)

September 23rd   June 22nd   December 22nd   March 21st

Fit the above dates to the following:

A   Spring Equinox   Summer Solstice

B   Autumn Equinox   Winter Solstice

C   Name the 4 seasons.

D   Remember: Learn:
       60 seconds =1 minute
       60 minutes =1 hour   *The months of the year*
       24 hours   =1 day   Thirty days hath September,
        7 days   =1 week   April, June and November.
       365 days   =1 year   All the rest have thirty-one,
       and   Excepting February alone,
       in one year there are   Which has twenty-eight days clear,
           52 weeks   But twenty-nine in each leap year.
           12 Calendar months
           13 Lunar months
       and 366 days in one leap year

Which of these is correct?

E   The earth rotates on its own axis during each
       24 hours   12 hours   365 days   7 days

F   The orbit of the earth round the sun takes
       365 days $365\frac{1}{4}$ days $365\frac{1}{2}$ days   366 days

Write answers only. Work across the page:

A   1 minute=   seconds   30 seconds =   minute

B   1 hour   =   minutes   1 day   =   hours

C   24 hours =   day   30 minutes =   hour

D $\frac{1}{4}$ min.   =   seconds $\frac{1}{4}$ hour   =   minutes

E   90 mins. =   hours $1\frac{1}{2}$ min.   =   seconds

F   1 week   =   days   A fortnight=   days

G   12 hours =   day   48 hours   =   days

H   15 secs.   =   minute   90 secs.   =   minutes

I   2 mins.   =   seconds   15 mins.   =   hour

J $1\frac{1}{2}$ hrs.   =   minutes   120 mins.   =   hours

K   2 days   =   hours   05 05 hours=   a.m.

L   8.5 p.m. =   hours   14 days   =   weeks

M   365 days =   year   366 days   =

N   1968 was a   year. A leap year is every   years

O   "Sept." stands for   "Jan." stands for

P   "Feb." stands for   "Oct." stands for

Q   June has   days   March has   days

R   Jan. has   days   Nov. has   days

S   Aug. has   days   May has   days

T   Feb. 1968 had   days   Feb. 1970 had   days

U   Feb. 1980 should have   days   Midnight is   hours

			Further practice pages
How many months from:			
A	April to Aug.?	July to Feb.?	65 A–F
Write in figures:			
B	5th Jan. 1972	26th Oct. 1970	65 G–I
If 1st July was on Sunday give the date of:			
C	The next Wednesday	The previous Monday	65 J–O
State how many days there are from:			
D	30th Jan. to 3rd. Feb.	25th Oct. to 1st Dec.	
E	24th Dec. 1970 to 12th Jan. 1971		65 P–T
F	26th Feb. 1972 to 15th Mar. 1972		
Write in figures:			
G	Twenty minutes past two in the afternoon		66 A–W
H	Twenty minutes to one in the morning		
Give the time 24 hours before:			
I	5.15 a.m. Friday		67 A–J
	6 p.m. 1st Jan. 1973		
Give the time 12 hours before:			
J	7.30 a.m. 30th June		67 K–O
	4 a.m. 1st Mar. 1972		
State how long it is from:			
K	11.40 a.m. to 12.15 p.m. mins.		
	7 a.m. to 5.45 p.m. hrs. mins.		67 P–T
L	7.30 a.m. Thursday to 10.15 a.m. Friday hrs. mins.		
Change these 24 hour clock times to 12 hour clock times:			
M	15 00 22 15 03 05		68 and 69
Change these ordinary times to 24 hour clock times:			
N	5.8 p.m. 7.20 a.m. 10 past m'night		70 A–I
State how long it is from			
O	10 45 to 14 28	22 35 to 01 10	70 J–U

Which month comes after:

A   Jan.   May   March   Aug.

B   Nov.   Oct.   Dec.   June

Which month comes before:

C   July   Feb.   Nov.   Jan.

Name the month which is three months after:

D   June 1970   Dec. 1970   Nov. 1971

Say how many months there are from:

E   Jan. to Mar.   months   Feb. to Sept.   months
F   Oct. 1970 to Mar. 1971   months   Aug. 1972 to April 1973   months

If "5.2.71" stands for "5th Feb. 1971" write what is short for:

G   3rd Jan. 1970   6th June 1969   17th Feb. 1970
H   22nd Aug. 1971   20 April 1971   25th Dec. 1972
I   31st Mar. 1972   1st July 1970   19th Oct. 1971

State which day is:

J   Two days before Sun.   Four days after Tues.
K   Three days after Sat.   Five days before Wed.

If the 12th June was on Wed., on which day was:

L   14th June   11th June   19th June   5th June

If 1st Aug. was on Mon. give the date of:

M   The next Mon.   The previous Mon.
N   The following Fri.   The previous Sat.

If 30th Nov. was on Sat., on which day was:

O   1st Dec.   29th Nov.   4th Dec.   23rd Nov.

Which are Leap Years?

P   1980   1982   1972   1970   1900   2 000

If 28th Feb. 1968 was on a Tues., on which day was:

Q   27.2.68   25.2.68   1.3.68   6.3.68

Write how many days there are from:

R   29th Jan. to 2nd Feb.   days   29th May to 2nd June   days
S   20th June to 5th July   days   6th Mar. to 6th April   days
T   20th Dec. '70 to 20th Jan. '71   days   5th Feb. '72 to 5th Mar. '72   days

Write in figures:

A   ten minutes past five o'clock

B   twenty minutes past six o'clock

C   half-past eight o'clock in the morning

D   half-past seven o'clock in the evening

E   twenty-five minutes past one o'clock in the afternoon

F   fifteen minutes past twelve o'clock mid-day

G   twenty-seven minutes past twelve o'clock mid-day

H   five minutes past twelve o'clock mid-night

I   half-past twelve o'clock mid-night

J   five minutes to four o'clock in the afternoon

K   five minutes to six o'clock in the afternoon

L   ten minutes to eight o'clock in the evening

M   twenty minutes to ten o'clock in the evening

N   twelve minutes to three o'clock in the afternoon

O   ten minutes to nine o'clock in the morning

P   fifteen minutes to five o'clock in the morning

Q   a quarter-past four o'clock in the morning

R   a quarter to four o'clock in the morning

S   a quarter to eleven o'clock in the evening

T   twenty-five minutes to twelve o'clock at noon

U   twenty-five minutes to twelve o'clock at mid-night

V   a quarter to one o'clock in the afternoon

W   eighteen minutes to one o'clock in the morning

Give the time two hours after:

A	6 p.m.	8 a.m.	12 noon	12 midnight
B	11 a.m.	11 p.m.	12.35 a.m.	11.22 p.m.

Give the time one hour before:

C	10.10 a.m.	11.44 p.m.	12 noon
D	1 a.m.	12.10 a.m.	12.35 a.m.

Give the time 24 hours after:

E   10 a.m. 7th June   12 midnight 1st Dec.

F   3.30 a.m. 30th April
      7.23 a.m. 28th Feb. 1971

G   8 p.m. 31st Dec. 1970
      2 a.m. 31st Dec. 1972

Give the time 24 hours before:

H   10 a.m. Sat.   12 noon 31st Mar.

I   12 midnight 2nd Jan.
     5 a.m. 1st July 1973

J   12 noon 1st Jan. 1971
     4 a.m. 1st Mar. 1972

Add one half-hour to:

K   9.15 a.m.   3.42 p.m.   11.30 a.m.
                        12.53 p.m.

L   11.45 a.m. Mon.   12 noon Tues.
          11.55 p.m. Fri.

Give the time 12 hours before:

M   6 p.m. Mon.   3.30 p.m. Sat.
         12 midnight Fri.

N   12 noon Tues.   12 noon Sun.
        8 p.m. 2nd July

O   8 a.m. 1st June   11.18 a.m. 1st May
      2.45 a.m. 1st Dec.

State how long it is from:

P	9.15 p.m. to 9.50 p.m.	7.30 a.m. to 8.10 a.m.
Q	11.30 a.m. to 12.10 p.m.	11.45 a.m. to 12.15 p.m.
R	12.15 p.m. to 2.20 p.m.	10.30 a.m. to 1 p.m.
S	9.50 a.m. to 12.57 p.m.	8.45 p.m. to 4.30 a.m.
T	9 p.m. Mon. to 7 a.m. Tues.	11.47 a.m. to 12.51 p.m.
U	8.30 a.m. Thur. to 9 a.m. Fri.	

In the past the 12 hour clock time has been used, but it is gradually being replaced by the 24 hour clock time, especially on railway and aeroplane time-tables.

The old way of telling time:

The new way of telling time:

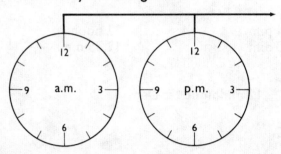
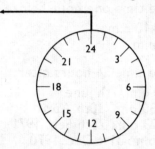

Here is a part of Mark's day as it would be on a 24 hour clock face:

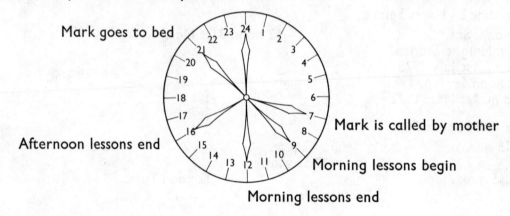

Mark goes to bed

Afternoon lessons end

Mark is called by mother

Morning lessons begin

Morning lessons end

No doubt you can see that such a clock face can cause confusion, for a half hour would be shown when the minute hand was a quarter of the way round, and the first hour would be shown when the minute hand was half way round the clock face.

We state the 24 hour time by adding 12 hours on to the 12 hour clock time, but many watches and clocks have faces like this:

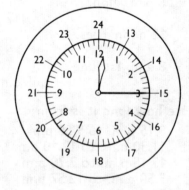

Instead of lessons ending at 4 p.m. we write 16 00 hours, leaving a space between hours and minutes instead of a stop. 1.30 p.m. becomes 13 30 and 8.25 p.m. becomes 20 25.

Like we have a special way of writing pounds and new pence we have a special way of writing time by the 24 hour clock.

To avoid mistakes we use place-holder zeros to make up the figures to four. There are always four figures.

9 a.m. becomes 09 00   9 p.m. becomes 21 00
5 past 2 p.m. becomes 14 05   7.30 a.m. becomes 07 30

Here is what a part of an airways time-table might look like. Rewrite it as it would be for a twelve hour clock, not forgetting a.m. and p.m.

A

	Mon.	Tues.		Wed.
	FA 526	FA 305	FA 623	FA 410
d.	09 20	10 30	19 35	21 50
a.	10 05	11 15	20 20	22 40
d.	10 50	12 00	21 05	23 30
a.	16 00	17 05	02 15	04 30
d.	17 10	18 20	03 05	05 00
a.	21 15	22 30	07 20	09 05

B   State the numbers of the flights (FA) which begin in the forenoon.

C   State the number of flights which begin in the afternoon.

D   State the number of flights which arrive in the forenoon.

E   How long did the flight FA 305 take?

F   Was flight FA 623 a faster or a slower flight than FA 305?

Change these 24 hour clock times to the 12 hour clock times:

G   13 00   15 30   17 00   20 00

H   08 00   10 00   02 10   11 08

I   04 20   03 42   20 20   21 06

Change these ordinary times to those of the 24 hour clock:

A	2 a.m.	2 p.m.	6 p.m.	6 a.m.
B	4.20 a.m.	4.20 p.m.	10 a.m.	10 p.m.
C	7.35 a.m.	7.35 p.m.	12 noon	12 midnight
D	12.50 p.m.	12.50 a.m.	1.5 a.m.	2.8 p.m.
E	10 min. past 3 p.m.	20 min. to 4 a.m.		16 min. past 10 p.m.
F	16 min. to 11 a.m.	25 min. to 12 noon		27 min. past 11 p.m.
G	a quarter to 12 noon		half-past 12 midnight	
H	a quarter past 12 noon		a quarter to 12 midnight	
I	40 minutes past midnight		10 minutes past midnight	

State the time two hours after

J	06 42	14 33	11 40	10 05	20 00
K	01 40	23 00	22 15	22 50	22 05

State the time one hour before

L	10 50	04 05	14 00	12 55	24 00
M	02 00	02 50	01 50	01 30	01 05

State the time one half hour before

N	13 00	02 00	04 35	21 10	20 15
O	23 15	10 30	10 15	02 05	01 40
P	01 20	00 35	00 15	00 05	01 01

State how long it is from

Q	09 00 to 10 40	10 05 to 11 50
	19 30 to 20 05	
R	01 55 to 03 30	17 10 to 20 05
	18 38 to 21 06	
S	24 00 to 02 30	23 00 to 01 15
	23 10 to 00 40	
T	22 00 to 01 45	22 00 to 01 10
	22 52 to 01 00	
U	08 40 to 11 17	23 05 to 00 50

					Further practice
					Practice 1 and pages

Work across the page:

Addition:

A

min. sec.	min. sec.	min. sec.	min. sec.	72 G–I
1 10	23 37	5 35	14 43	
2 22	9 40	45	8 0	
2 34	18 6	7 50	20 38	

B

hr. min. sec.	hr. min. sec.	hr. min. sec.	73 A–B
3 18 30	9 20 18	17 36 20	
5 20 6	15 7 9	47 57	
2 9 27	48 53	20 30 49	

C

wk. days hr.	wk. days hr.	wk. days hr.	73 C–F
3 1 14	5 6 21	14 3 16	
2 0 12	4 10	6 5 3	
5 4 8	7 0 6	6 21	
3 0 18	6 22	19 0 18	

Subtraction:

D

min. sec.	min. sec.	hr. min.	hr. min.	74 A–C
12 0	42 23	7 30	14 0	
− 9 15	−39 48	−6 54	− 9 49	

E

hr. min. sec.	hr. min. sec.	hr. min. sec.	74 D–E
5 15 15	10 0 0	18 36 29	
−4 20 50	− 7 33 47	−15 45 45	

F

wk. days hr.	wk. days hr.	wk. days hr.	74 F–H
6 3 10	10 0 0	16 0 12	
−2 2 11	− 7 4 8	−12 0 18	

Practice 1

Work across the page:

Change to minutes

A   65 sec.=   min.   sec.   80 sec.=   min.   sec.   105 sec.=   min.   sec.
           120 sec.=   min.   sec.   134 sec.=   min.   sec.

B   1 hr. 8 min.=   min.   1 hr. 26 min.=   min.   1 hr. 40 min.=   min.
               1 hr. 48 min.=   min.

Change to hours:

C   70 min.=   hr.   min.   95 min.=   hr.   min.   110 min.=   hr.   min.
           127 min.=   hr.   min.   144 min.=   hr.   min.

D   1 day 6 hr.=   hr.   1 day 12 hr.=   hr.   1 day 17 hr.=   hr.
                   1 day 22 hr.=   hr.

Change to days:

E   30 hr.=   day   hr.   40 hr.=   day   hr.   48 hr.=   days   hr.
               55 hr.=   days   hr.   60 hr.=   days   hr.

F   72 hr.=   days   hr.   80 hr.=   days   hr.   1 wk. 2 days=   days
               1 wk. 5 days=   days

Addition:

G

min.	sec.	min.	sec.	min.	sec.	min.	sec.	min.	sec.
1	15	1	30	4	43	6	38	11	58
1	22	2	18	6	9	10	50	8	4
1	29	1	26	2	36	7	5	20	30

H

min.	sec.	min.	sec.	min.	sec.	min.	sec.	min.	sec.
13	27	20	33	17	43	38	10	26	53
16	40	9	4	6	20	15	36	7	40
8	0	6	8	10	7	8	20	5	8

I

hr.	min.	hr.	min.	hr.	min.	hr.	min.	hr.	min.
6	15	12	36	8	30	21	37	18	36
	45	9	48	10	6	9	20	20	0
7	8	10	4	6	52	15	4	7	48

Addition

A

hr.	min.	sec.		hr.	min.	sec.		hr.	min.	sec.		hr.	min.	sec.
2	15	30		3	27	24		5	18	8		7	35	10
6	20	13		5	30	8		3	9	30		8	20	37
3	12	28		2	28	17		6	47	6		5	28	26

B

hr.	min.	sec.		hr.	min.	sec.		hr.	min.	sec.		hr.	min.	sec.
8	47	15		9	53	30		12	30	40			58	45
6	8	30			48	56		10	8	25		15	20	0
7	20	0		5	0	20			45	0			46	55

C

days	hr.		days	hr.		days	hr.		days	hr.		days	hr.
3	8		4	14		5	20		6	18		8	9
2	11		2	7		7	8		3	9		1	20
5	6		6	10		1	7		5	7		6	7

D

days	hr.		days	hr.		days	hr.		days	hr.		days	hr.
7	15		8	16		6	18		10	22		13	8
3	8		4	0		12	20		8	6		5	20
6	10		2	20		8	7		12	17		12	9

E

wk.	days		wk.	days		wk.	days		wk.	days		wk.	days
4	3		3	4		16	3		5	6		7	5
2	4		5	0			5			4		28	0
2	5		4	6		4	2		17	0		5	6

F

wk.	days	hr.		wk.	days	hr.		wk.	days	hr.		wk.	days	hr.
2	3	10		3	5	20		14	6	12		18	5	20
5	4	12		4	0	12			5	18			6	12
3	1	8		6	0	12		15	0	20		7	4	22
4	2	16			4	15		3	0	8		26	6	18

A

min. sec.	min. sec.	min. sec.	min. sec.	min. sec.
7 30	6 10	8 15	7 0	10 0
−5 8	−4 50	−3 45	−5 30	− 6 53

B

min. sec.	min. sec.	min. sec.	min. sec.	min. sec.
8 10	19 0	23 30	32 48	45 52
−7 45	−18 27	−20 52	−19 54	−43 58

C

hr. min.	hr. min.	hr. min.	hr. min.	hr. min.
5 40	7 15	6 50	10 35	12 0
−3 40	−4 30	−5 56	− 8 48	− 9 52

D

hr. min. sec.	hr. min. sec.	hr. min. sec.	hr. min. sec.
6 10 20	5 15 30	8 30 25	13 53 47
−4 9 28	−4 20 45	−6 37 49	−10 53 58

E

hr. min. sec.	hr. min. sec.	hr. min. sec.	hr. min. sec.
12 0 0	14 28 36	15 0 0	17 10 47
− 9 30 45	−10 27 46	− 54 40	−14 49 50

F

days hr.	days hr.	days hr.	days hr.	days hr.
7 15	6 0	9 12	12 18	14 0
−5 16	−2 18	−7 14	−10 20	− 9 8

G

wk. days	wk. days	wk. days	wk. days	wk. days
5 0	7 2	8 4	10 1	12 3
−3 4	−6 5	−7 6	− 4 4	−10 6

H

wk. days hr.	wk. days hr.	wk. days hr.	wk. days hr.
3 2 15	4 0 12	7 4 16	10 6 0
−2 1 17	− 5 22	−6 6 0	− 9 5 21

Check your multiplication and division of time

Work across the page:

Multiply:

					Further practice	
					Practice	Pages
A	min. sec. 2 21 × 3	min. sec. 3 14 × 7	min. sec. 5 35 × 8		2	76 E–F
B	hr. min. sec. 3 36 8 × 9		hr. min. sec. 7 30 42 × 12		2	76 G–I
C	wk. days hr. 4 5 14 × 6		wk. days hr. 6 6 18 × 12		2	77 A–D
Divide: **D**	min. sec. 6)8 0	min. sec. 7)9 41	min. sec. 11)2 23		3	78 A–C
E	hr. min. sec. 9)4 0 0		hr. min. sec. 12)21 40 0		3	78 D
F	days hr. 8)21 0	days hr. 10)6 16	days hr. 11)7 12		3	78 E–F
G	wk. days hr. 7)20 4 16	wk. days hr. 12)8 0 0			3	78 G

Practice 2

Work across the page:

Change to minutes:

A 75 sec. =   min.   sec.   100 sec. =   min.   sec.
   135 sec. =   min.   sec.   200 sec. =   min.   sec.
                    258 sec. =   min.   sec.

Change to hours:

B 90 min. =   hr.   min.   154 min. =   hr.   min.
   220 min. =   hr.   min.   275 min. =   hr.   min.
                    400 min. =   hr.   min.

Change to days:

C 30 hr. =   day   hr.   50 hr. =   days   hr.   86 hr. =   days   hr.
          103 hr. =   days   hr.   220 hr. =   days   hr.

Change to weeks:

D 15 days =   wk.   day 57 days =   wk.   day 89 days =   wk.   days
   150 days =   wk.   days   192 days =   wk.   days

E
min. sec.	min. sec.	min. sec.	min. sec.	min. sec.
2 10	2 18	4 21	3 26	5 23
× 4	× 3	× 4	× 6	× 7

F
min. sec.	min. sec.	min. sec.	min. sec.	min. sec.
4 15	5 40	4 37	5 48	6 53
× 7	× 8	× 9	× 8	× 9

G
hr. min.	hr. min	hr. min.	hr. min.	hr. min.
4 21	5 35	6 40	5 38	8 38
× 6	× 8	× 9	× 11	× 12

H
hr. min.	hr. min.	hr. min.	hr. min.	hr. min.
5 9	7 40	7 50	13 20	15 48
× 9	× 10	× 9	× 11	× 12

I
hr. min. sec.	hr. min. sec.	hr. min. sec.	hr. min. sec.
3 5 15	2 16 22	5 30 25	7 40 50
× 7	× 9	× 11	× 12

Multiplication

A

days	hr.		days	hr.		days	hr.		days	hr.		days	hr.
2	12		2	9		4	10		4	11		6	8
×	4		×	6		×	7		×	8		×	9

B

days	hr.		days	hr.		days	hr.		days	hr.		days	hr.
5	6		7	14		6	16		8	20		7	19
×	10		×	9		×	12		×	11		×	12

C

wk.	days		wk.	days		wk.	days		wk.	days		wk.	days
3	3		4	4		5	3		7	6		10	5
×	6		×	7		×	9		×	11		×	12

D

wk.	days	hr.		wk.	days	hr.		wk.	days	hr.		wk.	days	hr.
3	4	10		3	5	14		5	6	12		8	6	20
×		6		×		8		×		9		×		12

Practice 3

Work across the page:

Change to seconds:

A
1 min. 35 sec.=	sec.	2 min. 14 sec.=	sec.
2 min. 48 sec.=	sec.	3 min. 27 sec.=	sec.

B
2 min. 56 sec.=	sec.	4 min. 35 sec.=	sec.
4 min. 56 sec.=	sec.	6 min. 54 sec.=	sec.

Change to minutes:

C
2 hr. 28 min.=	min.	3 hr. 45 min.=	min.
5 hr. 28 min.=	min.	6 hr. 57 min.=	min.

Change to hours:

D
1 day 22 hr.=	hr.	2 days 12 hr.=	hr.
3 days 10 hr.=	hr.	3 days 21 hr.=	hr.

E
4 days 14 hr.=	hr.	5 days 9 hr.=	hr.
6 days 18 hr.=	hr.	7 days 23 hr.=	hr.

Change to days:

F
3 wk. 3 days=	days	5 wk. 5 days=	days
7 wk. 2 days=	days	8 wk. 6 days=	days

Division

A

min. sec.	min. sec.	min. sec.	min. sec.
5)6 10	4)6 20	6)2 18	7)2 41

B

min. sec.	min. sec.	min. sec.	min. sec.
8)10 32	9)3 0	7)3 16	9)5 15

C

hr. min.	hr. min.	hr. min.	hr. min.
6)7 12	8)10 0	11)4 35	12)17 0

D

hr. min. sec.	hr. min. sec.	hr. min. sec.
6)20 0 0	10)8 45 30	12)30 55 0

E

days hr.	days hr.	days hr.	days hr.
6)8 0	8)5 0	10)16 16	11)18 12

F

days hr.	days hr.	days hr.	days hr.
7)4 9	12)21 0	9)30 0	12)8 12

G

wk. days hr.	wk. days hr.	wk. days hr.
8)20 0 0	9)6 0 0	12)49 0 0